Hafiz Adnan Ahmad Baig
Syed Hussain Haider Rizvi

Mapping of Earthquake Prone Areas

Earthquake and it's assessment

Anchor Academic
Publishing

Baig, Hafiz Adnan Ahmad; Rizvi, Syed Hussain Haider: Mapping of Earthquake Prone Areas: Earthquake and it's assessment, Hamburg, Anchor Academic Publishing 2014

Buch-ISBN: 978-3-95489-309-6
PDF-eBook-ISBN: 978-3-95489-809-1
Druck/Herstellung: Anchor Academic Publishing, Hamburg, 2014

Bibliografische Information der Deutschen Nationalbibliothek:
Die Deutsche Nationalbibliothek verzeichnet diese Publikation in der Deutschen Nationalbibliografie; detaillierte bibliografische Daten sind im Internet über http://dnb.d-nb.de abrufbar.

Bibliographical Information of the German National Library:
The German National Library lists this publication in the German National Bibliography. Detailed bibliographic data can be found at: http://dnb.d-nb.de

© Anchor Academic Publishing, Imprint der Diplomica Verlag GmbH
Hermannstal 119k, 22119 Hamburg
http://www.diplomica-verlag.de, Hamburg 2014
Printed in Germany

Abstract

Traditional approaches provide only limited opportunities for general analysis of tectonic structure. However, Geographic Information System provides unique opportunities for solving a wide spectrum of problems, related to analyzing earthquake phenomena. GIS can also provide an effective platform for data assortment, organization, and research with multidirectional data sets. In addition, GIS can also provide an effective solution for integrating different layers of information. In this research an interactive, diversified seismological, geological, and geographical digital database has been developed for Pakistan. Using this GIS database, a better understanding of the tectonics and crustal structure of the region is possible. With the help of this database we have complied new maps of seismological parameters, depth to moho, P_n velocity and P_g velocity. The developed GIS database would help us in natural hazard evaluation, seismic risk assessment, and understanding of earthquake occurrences.

Keywords: Tectonic plates, GIS, Moho depth, P_n velocity, P_g velocity

Table of Contents

List of Figures

1.0   Introduction

Seismology is the study of seismic waves which are used to measure the intensity of earthquakes. Seismic waves are the waves of energy caused by the movements of tectonic plates. Geographically, Pakistan is situated on Eurasian and Indian tectonic plates. The Northwest (NW) Himalayan Thrust, the continental collision between the Eurasian and Indo-Pak palates formed the mighty Himalayas. Its north-west front is the most active seismic zone in the world.

It is noticeable from the seismic events of Pakistan that seismicity of this area is associated with the both surface and blind faults. Further, the surface faults events show that fault segments especially the hinterland zone are more active.

The damping effect of thick Precambrian salt is the reason of lesser seismic activity in the parts of active deformational front (Salt Range and Bannu).Pakistan and its neighboring countries come in high frequency earthquake range. The most severe Makran earthquake of 1945 with a magnitude of 8.3, affected Pakistan worst and created many offshore islands along the Makran coast.On the basis of plate tectonic features, geological structures, orogenic history (age and nature of the deformation, magnetism and metamorphism) and lithoffacies, Pakistan may be subdivided into the following broad tectonic zones i.e. Indus Plate Form and Foredeep, East Balochistan Fold and Thrust Belt, Northwest Himalayan Fold and Thrust Belt, Kohistan Ladakh Magamtic Arc, Karakorm Block, Kakar Khorasan Flysch

Basin and Makran Accretionary Zone, Chaghi Magmatic Arc and Pakistani Offshore

To conduct our research with multidisciplinary data sets, we need a convenient platform for data collection and organization that we get from GIS.

One of the most important features of a geographic information system is the manipulation and analysis of both spatial (graphic) and non-spatial (non-graphic) data. Every seismological parameter contains necessary information such as active fault and strike-slip fault etc. Full integration of GIS is needed to perform a standard seismic routine. In this research, GIS technology will be applied to regional scale tectonic problems of Pakistan.

Our main objective is to facilitate and enhance the capability to accurately locate and evaluate seismic events in Pakistan. The purpose of this research is also to explain the crustal and upper mantle structures of the tectonic plate in Pakistan. Application of GIS in seismology will help us better understand the tectonics and the crustal structure of the region. This study will also be used in natural hazard evaluation, better understanding of the earthquake occurrences, and seismic risk assessment.

1.1 Justification of the Research

This study will help us explain the crustal and upper mantle structure of the tectonic plate in Pakistan and will be used in natural hazard evaluation, better understanding of earthquake occurrences, and seismic risk assessment.

1.2 Objectives

There are two main objectives of this research work as given below:

1. Evaluation of seismic events in Pakistan

2. Development of GIS database of fault regions and seismic activity

1.3 Thesis organization

The section 1 gives a short introduction of Seismology, GIS, and the research work. Section 2 describes basic concepts of geological and seismological terms. Literature review for the thesis work is given in section 3. The section 4 covers the brief description of the study area. In section 5 data acquisition and techniques of data processing are explained. Finally, results and discussion of the research is given section 6

2.0   Background Concepts

The Earth's structure is made up of three major parts: the crust, the mantle, and the core. The crust is the upper most Earth's layer, with a thickness of 5 to 10 km for the oceanic crust, and 30 to 50 km for the continental crust. The crust is differentiated into an oceanic portion, composed of denser rocks such as basalt, and a continental crust portion, composed of lighter rocks such as granite. The Earth's mantle is a 2,900 km thick shell of compressed and heated rock, beginning below the Earth's crust. The center of Earth is referred to as the core. Chemically the core is composed of a mixture of iron, nickel, and trace of other heavy metals. The core can be divided into two layers e.g. the inner and the outer core. The base of Earth's crust is formed of big hard rocks known as tectonic plates. These plates provide support to crust and ceiling to mantle. There are three most important types of Tectonic plate boundaries: Divergent boundaries, Convergent boundaries, and Transform boundaries. Divergent plate boundaries are locations where plates are moving away from one another. Convergent plate boundaries are locations where lithospheric plates are moving towards one another. Transform boundaries are where two plates are sliding horizontally one another. The divergence and convergence of huge plates releases tremendous amount of energy jolting the surface of Earth. Each and every plate moves independently with its own speed but can affect other. There are various faults caused by the collision of tectonic plates. A brief description of a fault and its types is as under.

2.1 Fault

A crack or area of ruptures between two rocks is known as a "fault". Faults can vary from few millimeters to thousands of kilometers in length. During an earthquake, the rock on one side of the fault suddenly slips with respect to the other (Barazangi, et al., 1996). There are five main types of faults as given below:

2.1.1 Active fault

An active fault is a fault which has moved repeatedly in recent geological time and has the potential for reactivation in the future. The use of the word active may give the impression that the fault is actually in motion at the present time. The term "active faults" does refer to presently moving faults and those that will move in the future (Camp, et al., 1992).

2.1.2 Thrust fault

Above the fault plane, it is a dip-slip where upper block moves up and over the lower block.(Chaimov, et al., 1990).

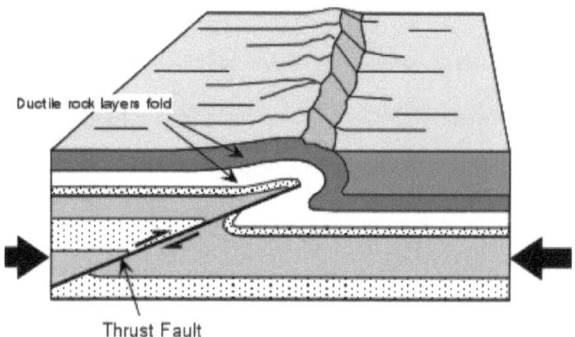

Figure 2. 1 Thrust fault (Image courtesy of Stephen Nelson, Tulane University)

2.1.3   Reverse fault

The mass of rock overlying a fault plane is known as hanging wall and the mass of rock lying below a fault plane is called foot wall. A reverse fault occurs when a hanging wall rises relative to the footwall. The areas suffering compression generate reverse faults (Sandvol, et al., 1996).

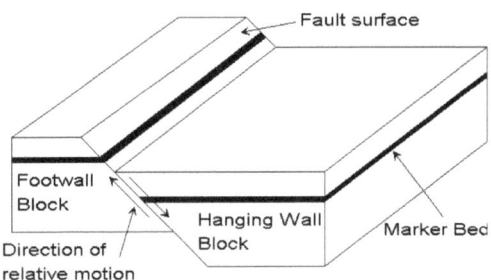

Figure 2. 2   Reverse fault

2.1.4   Strike-slip fault

The strike-slip faults are that type of faults in which the two slabs slip past one another. Strike-slip faults are subdivided as either right-lateral or left lateral. It depends upon whether the slip of the slab is to the right or the left. The slips take place adjacent to the strike, not up or down to the dip. In these faults the fault level surface is typically perpendicular; as a result there is no hanging wall or footwall. The force generating these faults is lateral or horizontal, moving the sides past each other (Le Pichon, et al., 1978).

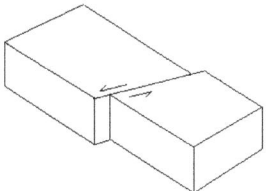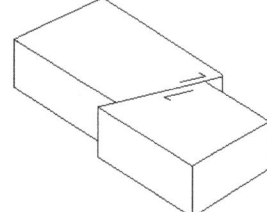

Figure 2. 3 Strike-Slip faults

2.1.5 Lineaments

Crustal lineaments are major 'fundamental' faults or fault zones which have had a lasting influence on the geological evolution of the continental lithosphere. They may be generated by a single strand of intense deformation or fracturing, or may consist of a complex geometrical array of faults and shear zones (Dewey, et al., 1979).

2.2 Seismological parameters

Seismological parameters are used to map the Earth's interior and to study its physical properties. The extension of the Earth's shallow crust, deeper mantle, liquid outer core, and solid inner core are inferred from variation in seismic velocity with depth. Most of the information about the nature of the fault is determined from the results of seismograph. The seismological parameters are Moho depth, P_g velocity and P_n velocity. A brief description of these parameters is given below:

2.2.1   Moho

Moho is the boundary between the Earth's crust and its mantle. Mostly Moho lies at a depth of about 22 mi (35 km) below continents and about 4.5 mi (7 km) beneath the oceanic crust (Ghalib, 1992). It is revealed by the latest instruments that the velocity of the seismic waves increases swiftly in Earth's Crust.

2.2.2 P_g and P_n velocity

The propagation of the seismic wave through the Earth's interior is governed by the law of light wave in optics. If the propagating velocities and other elastic properties were uniform through the Earth, seismic wave would radiate from the focus of the earthquake in all directions through the Earth along a rectilinear path. The velocity of the seismic wave in the granitic layer is called P_g velocity. Where as its velocity in the basaltic layer is called P_n velocity.

3.0 Literature Review

Remote sensing and GIS have been used to extract the spatial distribution of faults and other geologic structures, to study and interpret the active tectonics in South West Pakistan and elsewhere. Faults are natural simple or composite, and linear or curvilinear features, perceptible on the Earth's surface, which may depict crustal structure or represent a zone of structural weakness (Masoud and Koike, 2006). The strains that initiate from stress concentration around flaws, heterogeneities, and physical discontinuities, mainly appear in the form of faults, fractures and joint sets, originate them. (O'learly, et al., 1976; Davis, 1984; Clark and Wilson, 1994).

Yun and Moon (2001) proposed a faults extraction technique from Digital Elevation Model (DEM) using drainage network, which may relate to the lineaments of the underlying bedrocks. North and Pairman (2001) proposed a smoothening filter for remotely sensed imageries to detect edge boundary between two different land cover objects in North West Frontier Province (NWFP) and Kashmir. Leech et al. (2003) used digitally processed Landsat TM imageries to identify the faults in the coastal and northern areas of pakistan, and successfully interpreted the kinematics of the area by analyzing the statistics of faults frequency and their spatial distribution. Nama (2004) used Landsat Enhanced Thematic Mapper (ETM) to detect newly formed lineaments due to plate movements in Mardan, Swabi and Buner. Ali and Pirasteh (2004) used digitally processed Landsat ETM imageries for mapping and structural interpretation in the Zagros structural belt. They concluded that remote sensing can be very helpful to detect new geologic structures and to confirm previously field-mapped faults and folds. Jansson and Glasser (2005) found that False Color Composite (FCC) images created by combining Thermal Infra-Red (TIR) and Near Infra-Red (NIR) bands of Landsat ETM+ draped over the Digital Terrain Model (DTM) substantially enhanced the lineaments identification. Mostafa and Bishta (2005) used Landsat ETM+ imageries to calculate the strike slip fault density map of Balochistan in Pakistan, and correlated the lineament density with radiometric map, and located new uranium targets. Abarca (2006) proposed a semi-automatic technique, called Hough Transformation (HT), to extract linear features from grid based DEM in salt range region, and found that it was one of the most efficient and time economic ways to detect linear features like fold and fault. Masoud and Koike (2006) used Landsat ETM+ imageries and Digital Elevation Model, obtained from the Shuttle Radar Topographic Mission (SRTM-DEM), to analyze the spatial variation in the orientation of the thrust fault, and correlated them to the geology and hydrogeology of the Kashmir. Hearn (1999) used remote sensing to study active faulting and folding by observing and analyzing the geomorphologic features in southern Kerman province, North-West of the Makran accretionary prism.

Many authors have studied the structure, tectonics, and mechanism of latest deformation in the Makran accretionary prism in Iran and Pakistan. Hawkins (1974) has identified multiple co-existence of subductions: Mesozoic subduction, characterized by blueschist, quartzite, and marble, conserved south of the Jaz Murian Depression, Cenozoic subduction, characterized by the presence of calc-alkaline intrusions north of the the Jaz Murian Depression. Seismic study by Eide, at el., (2002), that found a series of low velocity zones within the accretionary wedge, suggests thrusting of compacted older sediments over younger ones, or presence of a large amount of fluid expulsion towards north of Pakistan. Landward flow of a large amount of fluid, expelled by subduction, is also evident by the presence of mud diapirs and mud volcanoes that occur in the accretionary complexes of Iran and Pakistan (Tahirkheli, et al., (1979). Yeats, et al., (1984) used swath bathymetric images and seismic reflection data to study the evolution and deformation of submarine convergent wedges in the Makran accretionary wedges of Pakistan. Field work done by Smith et al. (2005) suggests that the decreasing intensity of East West trending folds and thrusts, from north to south across the prism, expresses a bulk North-South Eocene to Miocene shortening in Iran.

All the preceding studies explained above are based on the conventional method of Geology and advanced techniques of remote sensing. But present study is a combination of the Geographic Information System (GIS) and seismology. Gathering geological data and disseminating the data by conventional methods is a dawdling and costly procedure. In order to re-address the insufficiency of world wide geological information, the more rapid approaches are needed within a reasonable time frame and cost. In modern age various affordable operational technologies such as GIS can significantly contribute to improve efficiency. GIS has been shown a significant help in the study of seismic hazard and risk. These efforts have consequenced in a considerable expansion of knowledge in seismological analysis (Dhakal, et al., 2000).

4.0   A Brief Description of Study Area

Pakistan is located in South Asia and has an entire area of 803,940 square kilometers. Pakistan is bordered by India to its east and shares 2912 km long border with India. Iran is in the west with a 909 km boundary. Afghanistan is in the northwest having a Durand line of 2430 km. In Northeast, the great Himalayas create a 523 km long wall with China. Arabian Sea is in the west providing Pakistan with a 1046 km long coastal line.

Pakistan is an elongated territory between the Arabian Sea and Karakoram peaks, exist obliquely between 24° to 37° North latitudes and 61° to 75° East longitudes. Topographically, Pakistan has a continuous massive mountainous region in the north, the west and south-west and a huge fertile plane, the Indus plane. The northern mountain structure, comprising the Karakoram, the great Himalayas, and the Hindu-Kush, has massive mass of snow and glaciers, and hundreds of peaks.

Geologically, Pakistan is situated on Eurasian and Indian tectonic plates. The Figure 4.1 shows some stages of the drift of the Indo-Pak Subcontinent in Tethys Sea. Between 1897 and 1952 there was a time of extremely more seismicity when 14 major earthquakes (Magnitude ≥ 7.5) occurred which also includes 5 great earthquakes of Magnitude ≥ 8. Distribution of seismic events within Pakistan indicates that seismicity (Magnitude ≥ 4.0) appears to be associated with both the blind and surface cracks (Sengor, et al., 1985). At the same time, appearance of more events near the surface faults indicates that some fault sections, especially in the hinterland zone, are more dynamic. In parts of the energetic deformational front (for example Salt Range and

12

Bannu) less important seismic events (Magnitude ≥ 4.0) may possibly exist because of the damping effect.(Armbruster, et al., 1978)

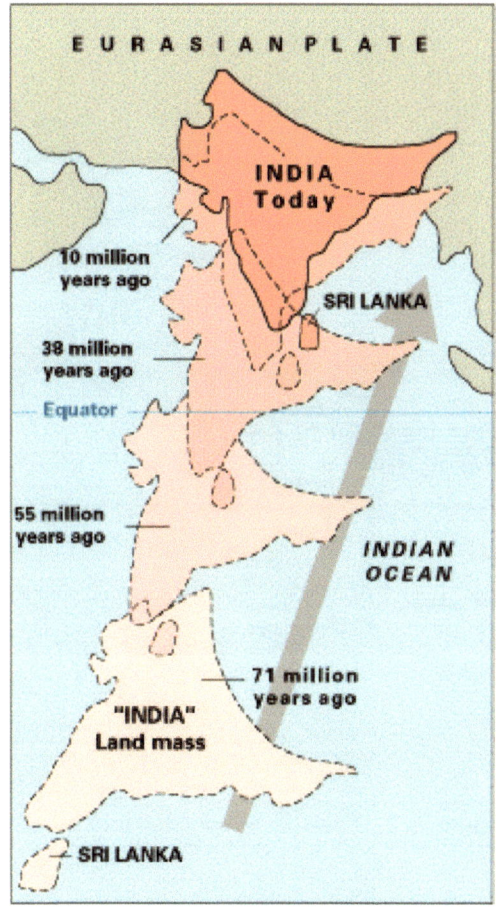

Figure 4. 1 Stages of the drift of the Indo-Pak subcontinent in Tethys sea (Source: Powell, 1979).

On the basis of plate tectonic features, geological structures, orogenic history (age and nature of the deformation, magnetism and metamorphism) and lithoffacies, Pakistan may be subdivided into the following broad tectonic zone. These zones are also shown in Figure 4.2

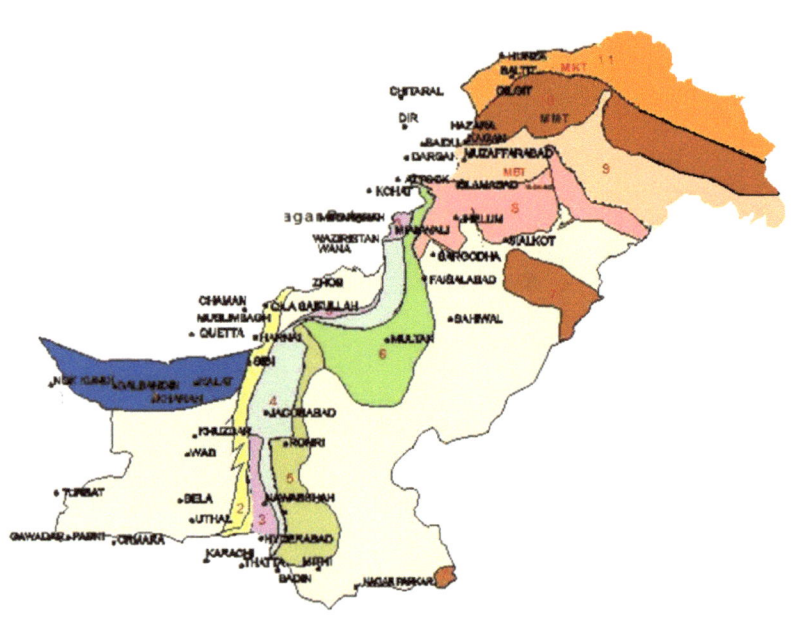

Figure 4. 2   Pakistan subdivided into broad tectonic zones

- Indus Plate Form and Foredeep

- East Balochistan Fold and Thrust Belt.

- Northwest Himalayan Fold and Thrust Belt.

- Kohistan Ladakh Magamtic Arc.

- Karakorm Block

- Kakar Khorasan Flysch Basin and Makran Accretionary Zone.

- Chaghi Magmatic Arc

- Pakistani Offshore

Indus Plate Form and Foredeep zone extends over an area exceeding 250.000 km in the south eastern Pakistan and includes the Indus Plane and Thar Cholistan Deserts.It hosts more than 80 % of Pakistan's population.

East Balochistan Fold and Thrust Belt zone of folds and thrust is 60 to 150 km wide, with a strike length of about 1,250-km. It extends southward from Waziristan, through Loralai-Bugti and around the Quetta syntax's down south to Karachi and Indus delta. The East Balochistan Fold and Thrust Belt are the product of transpression and oblique collision of India-Pakistan plate with the Afghan block (Burget, al., 1996). This belt is reportedly underlain by relatively thinner transnational or oceanic crust at least in northern part of Balochistan (Edwards, et al., 2000). Towards the west, the outer part of the East Balochistan Fold and Thrust Belt is comprised of an over 550-km long and 20 to 40 km wide imbricate zones of thrusts and nappes, with melanges wedge (Jadoon, et al., 1994).

The Northwest Himalayan Fold and Thrust Belt occupies a 250 km wide and about 560 km long area. This Irregular shaped mountainous region stretches from the Afghan border near Parachinar, up to the Kashmir Basin .The Hazra Kashmir and Nanga Parbat Syntax's forms its eastern margin (Johnson et al., 1985). It covers all the terrain between the Main Mantle Thrust (MMT) in the north and Salt Range Thrust in the south. This region is comprised of the mountain ranges of Nanga Parbet, Hazara, Southern Kohistan, Swat, Margalla, Kalachitta, Kohat, Sufaid Koh, Salt Range and its western extension (Pognante, et al., 1991).

Kohistan Ladakh Magamtic Arc is an intraoceanic island arc bounded by the Indus Suture zone (Main Mantal Thrust or MMT) to the south and the Shyok Suture zone (Main Karakorm Thrust or MKT) to the north. This East-West oriented arc is wedged between the northern promontory of the Indo Pakistan crustal plate. Karakorm block is 70 to 120 km wide and 1,400 km long structural zone (Auden, 1938)

Kakar Khorasan Flysch Basin And Makran Accretionary Zone are two tectonic zones presently form separate and distinct structural units northwest and west of the Sulaiman-Kirther fold and thrust belt (Vince, et al., 1996).

Chaghi Magmatic Arc extends north of the Kharan depression, the Chaghi arc is an East West trending arcuate, south-verging magmatic belt comprised of cretaceous to tertiary and volcanic (Petterson,et al., 1996).

The Pakistani Offshore extends for 700km from Rann of Cutch to the Iranian border (near Gwadar). It comprises two distinct structural and sedimentary basins (Deslo, 1930). The Indus and Makran Offshore Basins which are separated by Murray Ridge.This ridge is an extension of the Owen Fracture

Zone and forms boundary between Indian and Arabian Plates (Searle, 1986).

5.0 Data and Processing

5.1 Data Acquisition

Three types of data sets have been used in this research. These data sets are geographical datasets, geological datasets and seismological datasets

The geographical data sets are international boundaries and district boundaries with their attributes. The district map of Pakistan has been obtained from the World Wide Fund for Nature (WWF). This Map is shown in Figure 5.1. The attributes of this map are the names of the districts with their corresponding provinces. The attribute table for the district map of Pakistan is given in Appendix A.

Geological data sets consist of tectonic plate boundaries and different types of fault lines of the region under study. We have acquired the seismotectonic map from the Pakistan Meteorological Department (PMD), Islamabad which is shown in Figure 5.2. This map has been used in this research for the study of various fault lines in Pakistan.

Seismological datasets consist of seismic parameters i.e. Moho depth, P_n velocity and P_g velocity. Seismological parameters have been obtained from the Regional Meteorological Centre Peshawar. We have available seismological parameters for 114 locations with their geographic coordinates. These seismological parameters have been calculated from the seismographs for the period January 1983 to June 2009. Moho depth, P_n velocity and P_g velocity datasets are given in appendix B, C and D respectively.

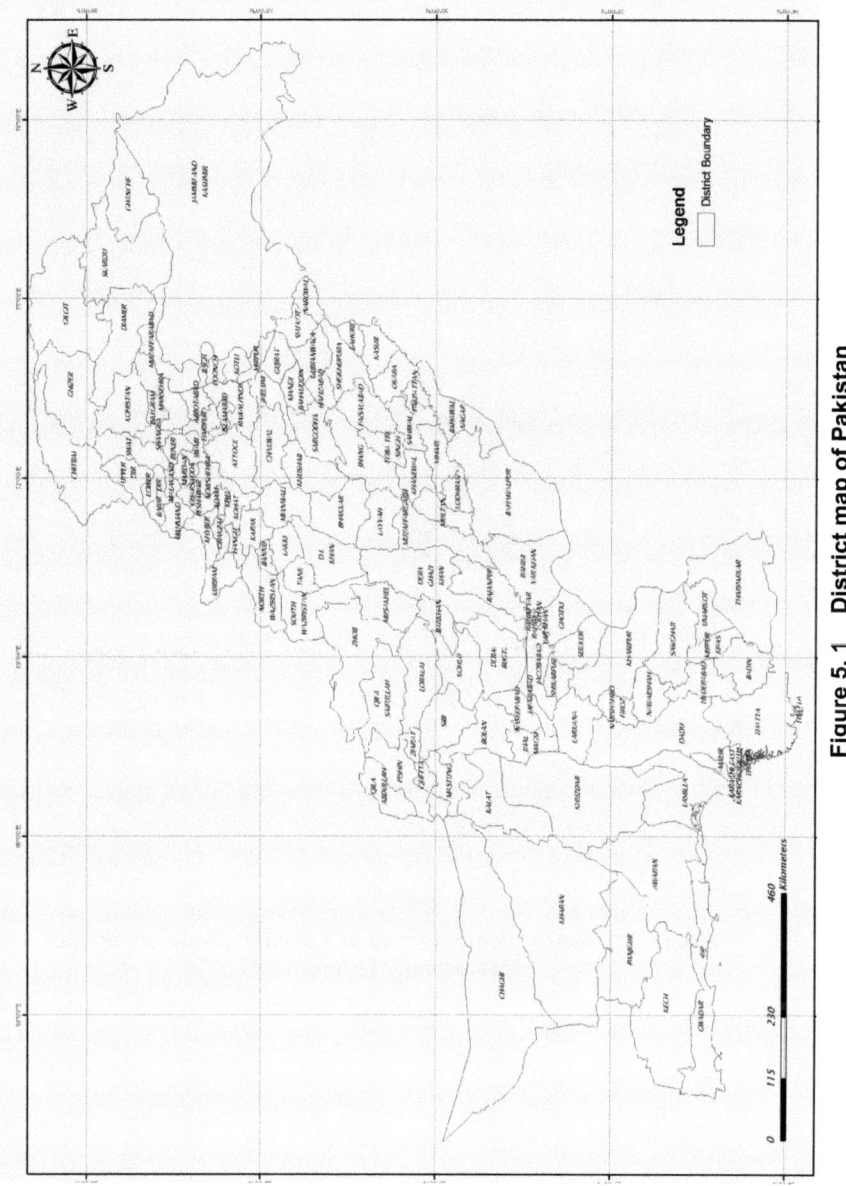

Figure 5. 1   District map of Pakistan

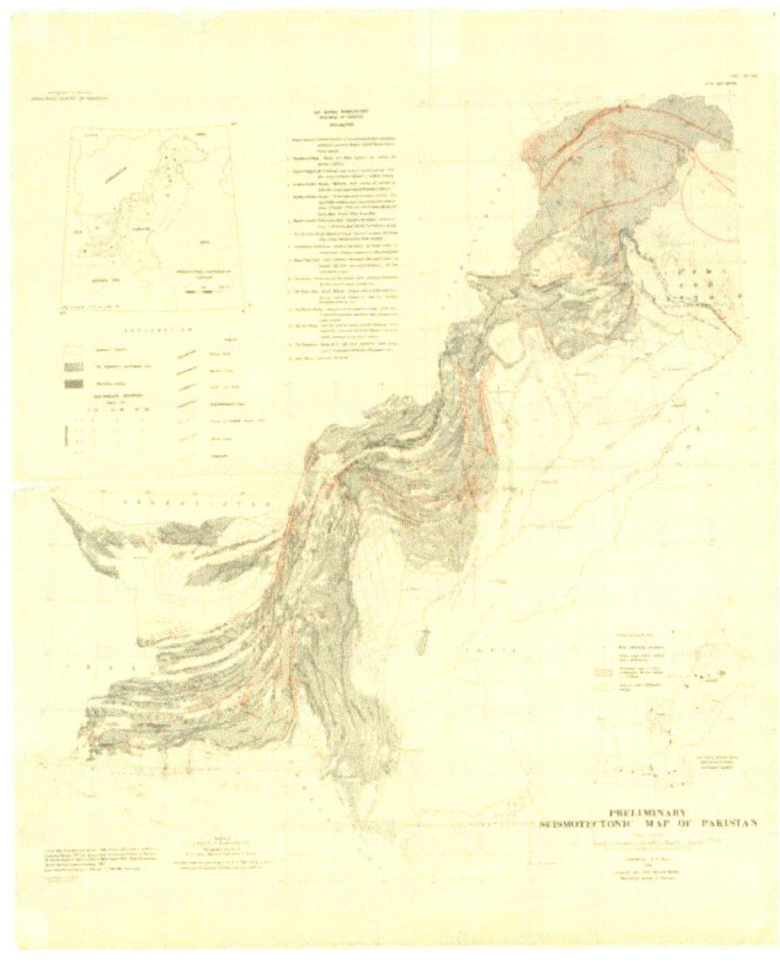

Figure 5. 2 Seismotectonic map of Pakistan (Source Pakistan Meteorology Department)

5.2   Processing of the Geographical and Seismological Data

The following techniques have been employed for the processing of geographical and seismological data used in this research.

Georeferencing is a method of allocating coordinates to points on the image of known locations that can also be easily differentiated on a paper map. Common methods for geo-referencing are affine transformation, geo-reregister method, and polynomial equations. We have two source maps such as seismotectonic map and district map of Pakistan. We have applied the Geo-reregister method for the geo-referencing of the seismotectonic map and district map of Pakistan. Geo-reregister method can geo-reference the seismotectonic map and district map of Pakistan through rotation, translation, and scaling. The Geo-reregister method is usually an iterative process. This method is run with an initial set of ground control points. We have required these ground control points for the geo-referencing of the seismotectonic map and district map of Pakistan. The ground control points are not part of a map and must come from other sources such as GPS or gridded maps. The ground control points have been selected from the grid overlaid upon these two maps. As the control points are identified, their real-world co-ordinates can then be obtained from grid source maps. The WGS 84 projection system is used in geo-referencing of seismotectonic map and district map of Pakistan.

Digitizing is the procedure of changing data from raster to vector format. Data sources for creating vector data include seismotectonic map and district map of Pakistan. Seismotectonic map and district map of Pakistan are primary data sources in this research work. The accuracy of digitizing is directly related to the accuracy of these source maps. In this research On-screen digitizing method has been used. This is manual digitizing on the computer monitor using a scanned copy of geo-referenced seismotectonic and district map of Pakistan as the back ground. On-screen digitizing is an efficient method for digitizing. We have employed this technique to create shape files of districts of Pakistan, active faults, thrust faults, reverse faults, strike-slip faults, and lineaments. The WGS 84 projection system has been used for shape files.

The next step is the plotting and interpolation of the seismological parameters such as P_n velocity, P_g velocity and Moho depth at different locations in Pakistan. We have the data of P_n velocity, P_g velocity and Moho depth in database file format (dbf). We have maneuvered these seismological parameters against their coordinates using Arc GIS 9.2 and have obtained a shape file for each parameter. We have also interpolated these seismological parameters using Inverse Distance Weighted (IDW) interpolation technique. The Inverse Distance Weighted (IDW) is referred to as deterministic interpolation technique as it directly depends upon the adjacent calculated values. By this interpolation technique we get values of seismological parameters for the entire region under study.

Finally we have produced maps of P_g velocity, P_n velocity, Moho depth, fault lines and seismic vulnerability in JPEG file format. In order to identify seismically vulnerable zones we have overlaid shape files of seismological parameters upon the district map of Pakistan. For the generation of final out put maps re-sampling process is required. In re-sampling map projection, re-projection, or geometric transformation creates a new grid. Data re-sampling is therefore required to fill each cell of the new grid with the value of the corresponding cell or cells in the original grid. Three common re-sampling methods for filling the new raster grid are nearest neighbor, bilinear interpolation, and cubic convolution. The nearest neighbor re-sampling method fills each cell of the new grid with the nearest cell value from the original grid. The bilinear interpolation method fills each cell of the new grid with the weighted average of the four nearest cell values from the original grid. The cubic convolution method fills each cell of the new grid with the weighted average of the 16 nearest cell values from the original grid. In this research work the nearest neighbor re-sampling method has been used.

5.3   Georeference a Topographic Map

Open Arc Map to a new empty map.

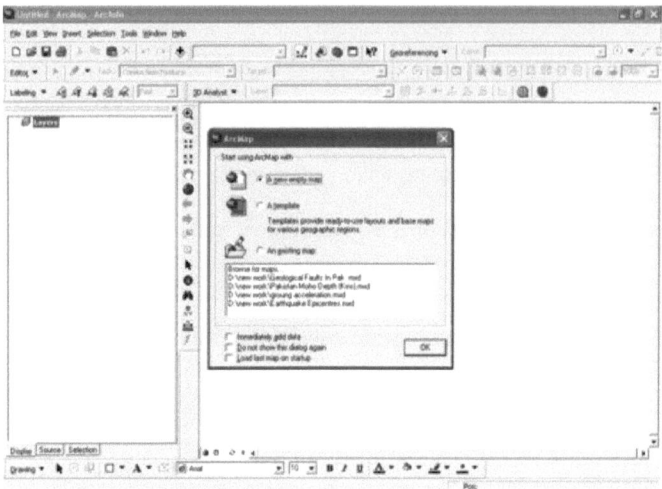

Using the Add data button add the map image you wish to georeference.

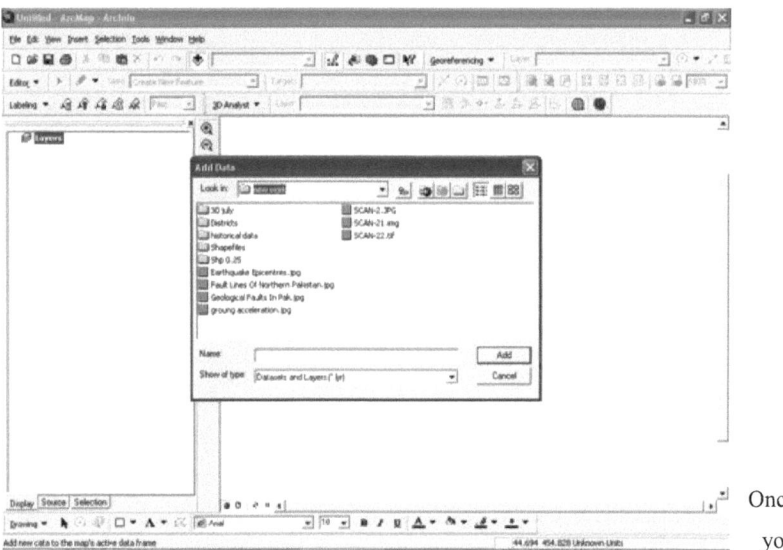

Once
you
click on the Add button, you will get the message about missing spatial reference information. Click OK in this dialog box; this will add the image to the data frame.

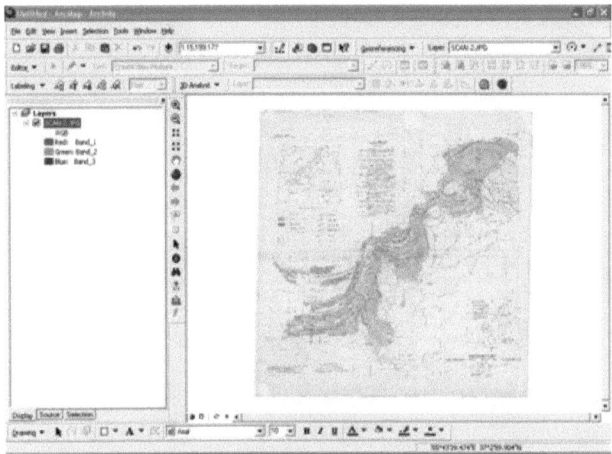

To view this new image you need to access, with a right-click on the newly added layer to get to the menu options, Zoom to Layer

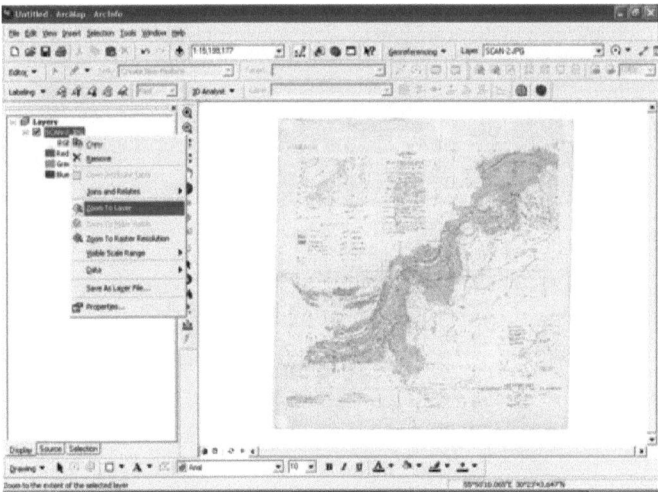

Click on the main menu bar and open the georeference tool bar

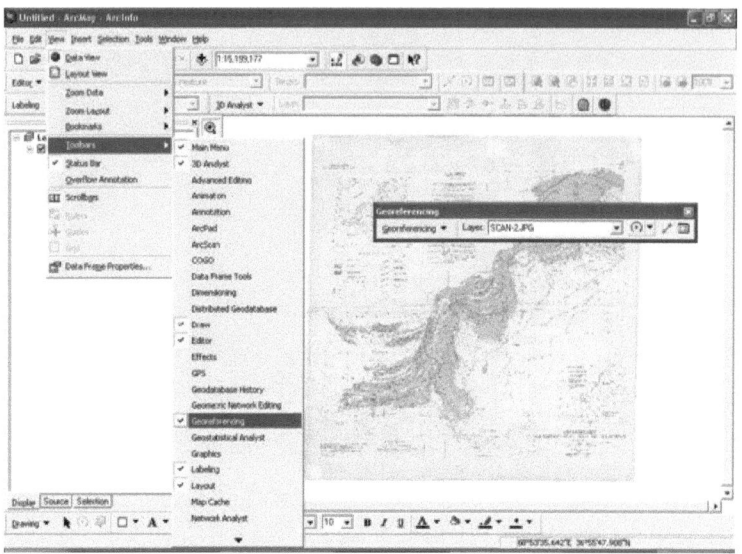

Zoom the upper left corner of the scan map. Now select the add control point button. Click on suitable area of the scan map to add the control point.

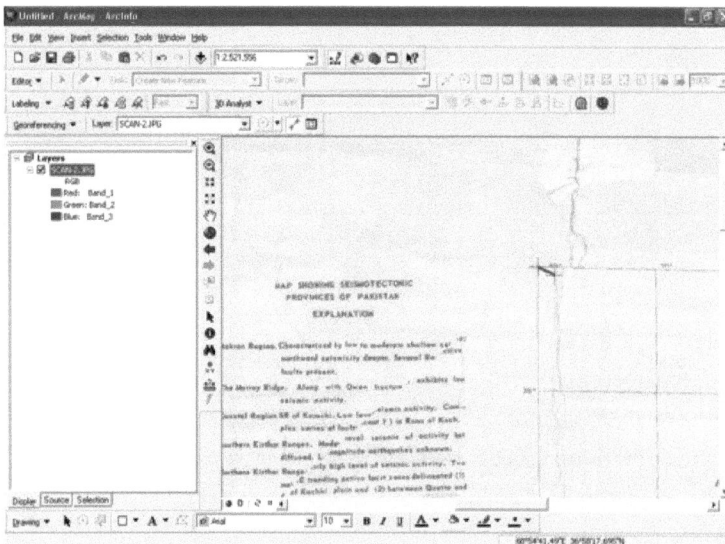

After the control point is added, open the View Link Table. Replace X and Y values only. Be consistent with the order you assign the control points and the order of your coordinates to avoid misplacement.

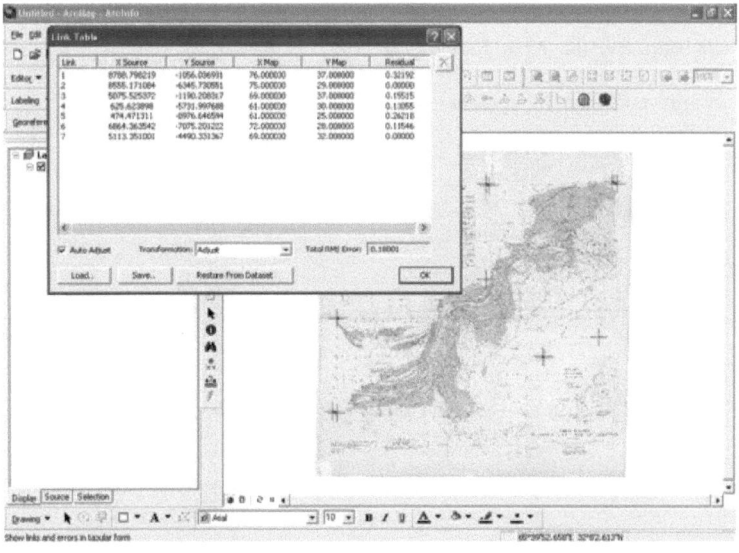

The degree to which the transformation can accurately map all control points can be measured mathematically by comparing the actual location of the map coordinate to the transformed position in the raster. The distance between these two points is known as the residual error. The total error is computed by taking the root mean square (RMS) sum of all the residuals to compute the RMS error. This value describes how consistent the transformation is between the different control points (links). Links can be removed if the error is particularly large, and more points can be added. While the RMS error is a good assessment of the accuracy of the transformation, do not confuse a low RMS error with an accurate registration.

Once all control point coordinates have been entered into the link table, use the drop-down menu on the georeference toolbar to access the Rectify command. Click on Rectify, the Save As dialog box will appear on the screen. This image shows the default cell size, change to the Resample Type to Bilinear, and a new filename for the

Output Raster with the path to the data folder. Once you click on save geoprocessing will start and the Auto Run dialog box will appear on your screen showing the progress and when it has completed rectifying your image.

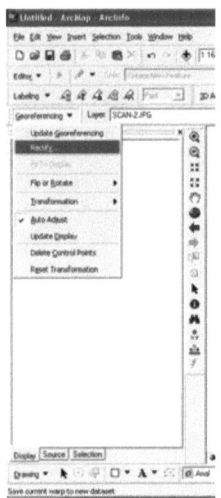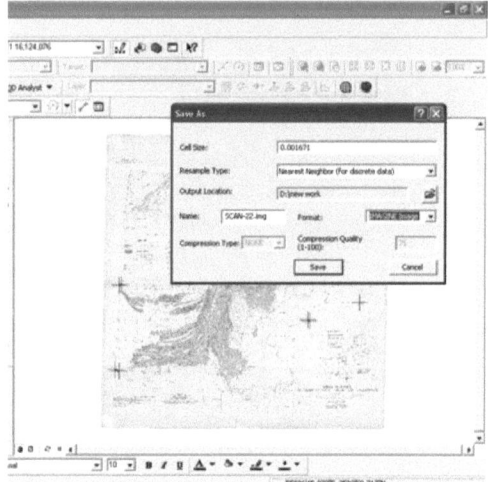

Now scan map is georeference and save this map in tiff format.

5.4   Digitization of Map

Start ArcMap, and create a new, blank document.

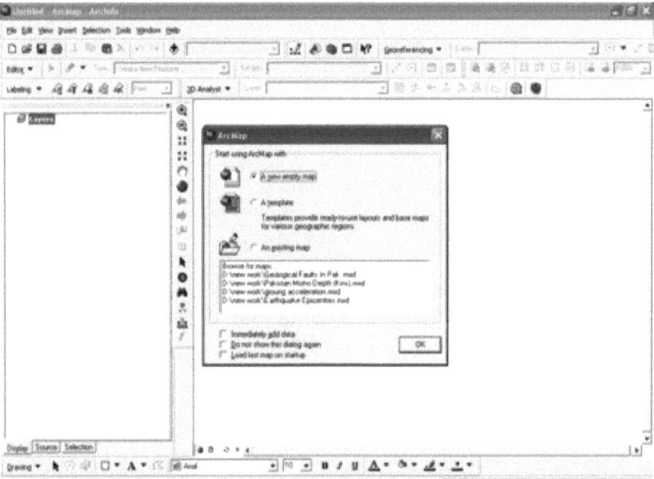

Select the add "Add Data" button; navigate the map wish to Digitize

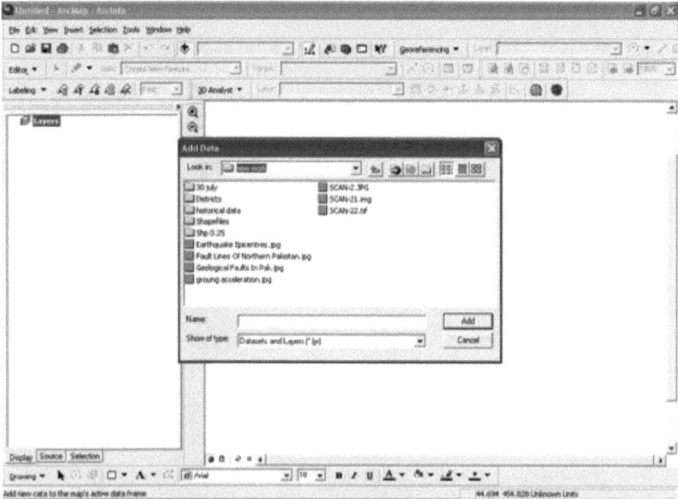

This is a scanned map often used for mapping. the coordinates change in the lower right corner of the ArcGIS window whenever move the mouse. These are the coordinates for the view, and they have been established from the map data.

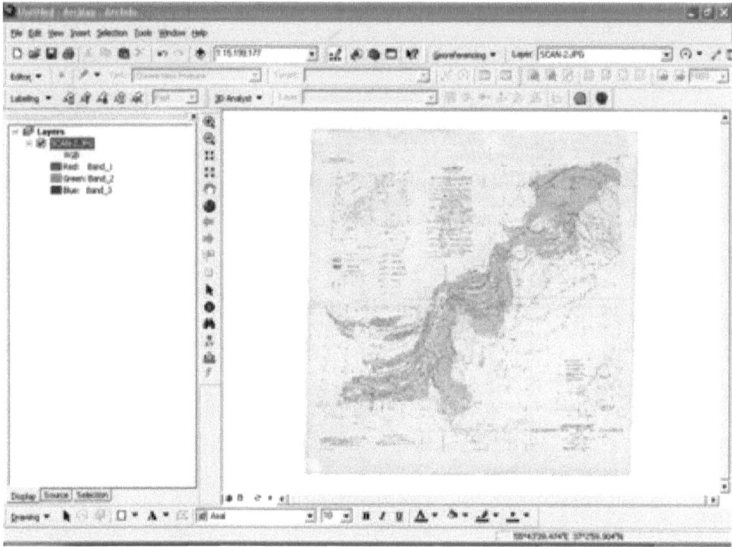

Open ArcCatalog by clicking on the icon in the main ArcMap toolbar:

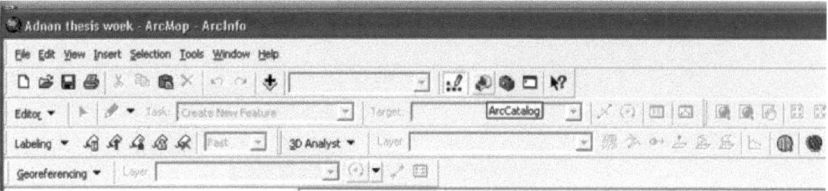

This should open a window with two panes, a navigation panel on the left, and a file display on the right.

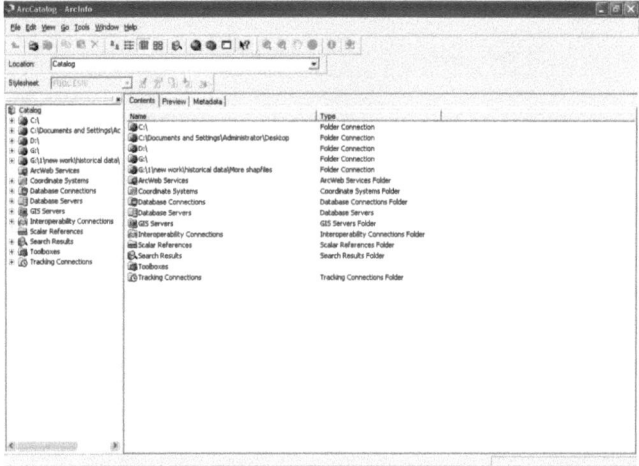

Left click on File at the top of the ArcCatalog window, then New, then Shapefile

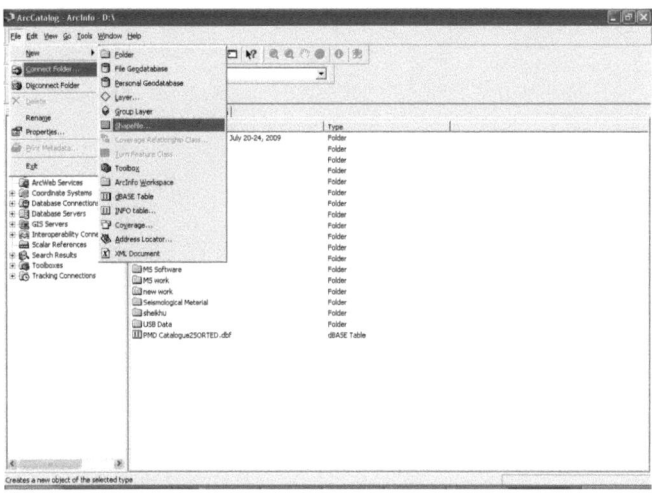

This should open a "Create New Shapefile" window.

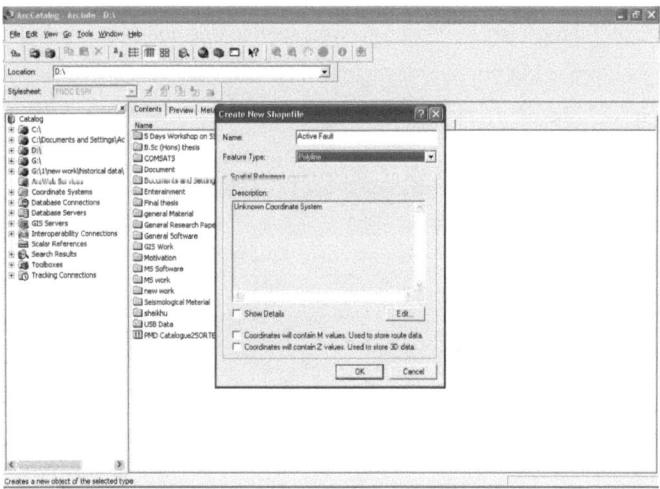

The default feature type will initially come up as point, but you could also select line or polygon as the default types, using the selection triangle to the right of the Feature type block. Enter something descriptive for a Name, e.g., Active fault. Specify a feature type of polyline. Left click on Edit to set the coordinate system. The resultant set of windows will be the projection windows

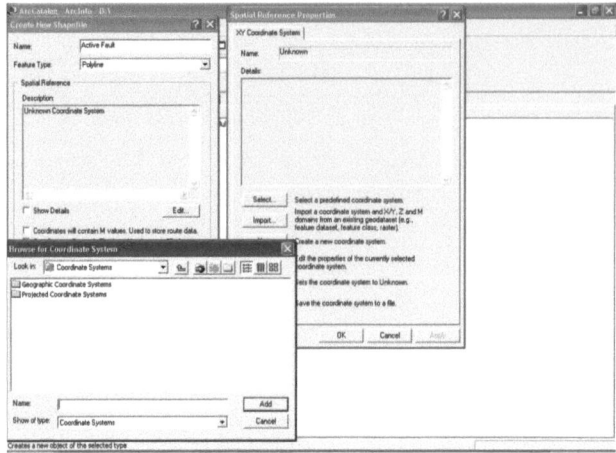

Specify Select > Projected Coordinate System > UTM > WGS84> Zone 42 N system also specify a coordinate system from an existing layer, via Edit.

The goal is to digitize points (earthquake locations), lines (fault lines), and polygons (International boundary) from this map.

Left click on Editor, Start Editing. After you start editing and select "active fault" as your target layer, select the layer to digitize. The active fault (or whatever you named the shape file) should appear as the Target of you edits. If not, use the selection triangle to specify the empty point file you created using ArcCatalog.

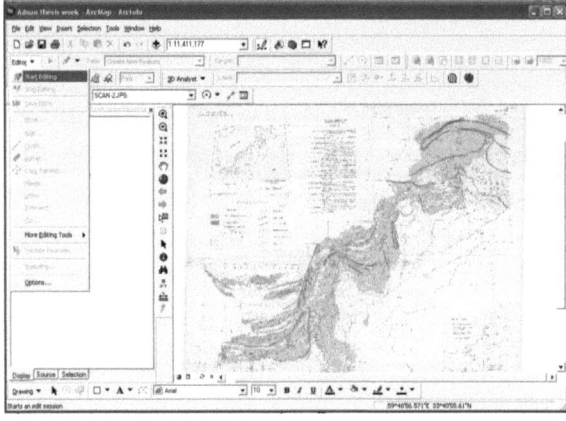

After the completion of digitizing, use Editor ---- Save Edits.

5.5   Maneuverings of Seismological Parameters

We have the data of Moho depth of Pakistan in database file format. This data is converted in shapefile for the creation of the mapping of Moho depth. Click on the tool and then "Add XY Data"

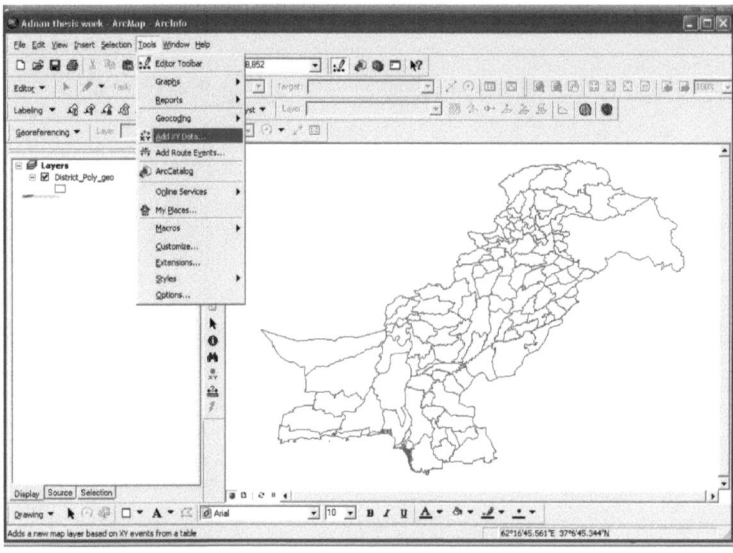

The new window is open in which add the database file.after selecting the projection system and coordinate system click "OK" points are added.

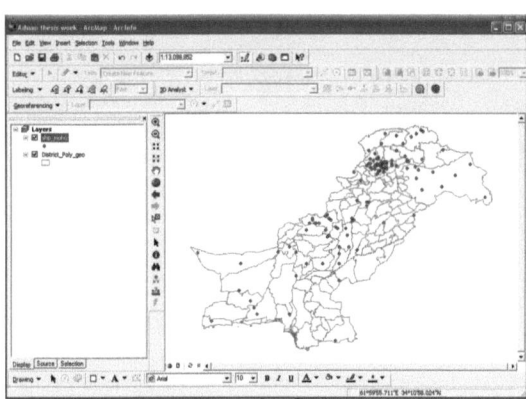

Right click on the add moho depth point file and then click on the export data the new window is open.

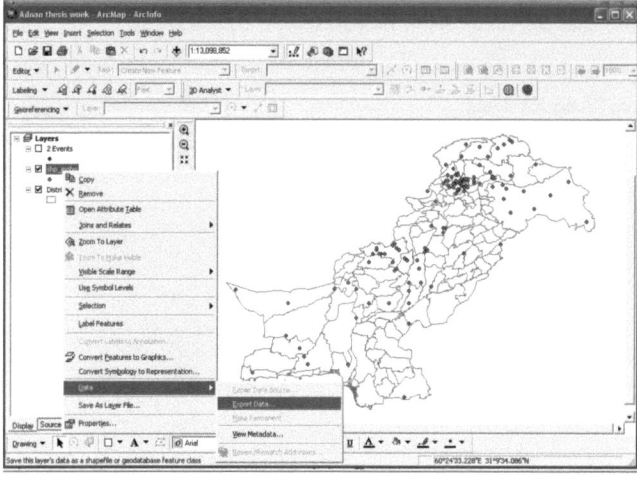

Give the name of the shapefile and then click "OK". The shape file of Moho depth of Pakistan is generating. The same process is for the Pn velocity shapfile and Pg velocity Shapefile.

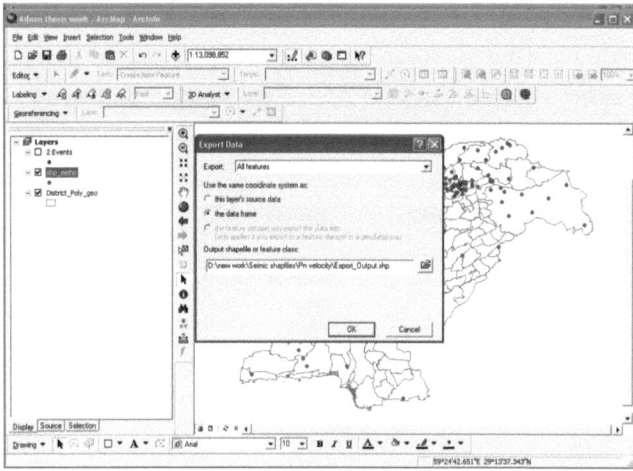

6.0 Results and Discussion

Now we are analyzing the resultant maps of geographical, geological and seismological data sets.

6.1 Analysis of Geological Faults Location in Pakistan

The Figure 6.1 shows a map of geological fault lines in districts of Pakistan. Different colored symbols in the map legend represent fault lines on the map. The symbol for district boundaries is also shown in the map legend. Names of the districts are clearly visible on the map. This map is a combination of geological and geographical data sets so we can say that the map is the GIS of geological fault locations in Pakistan. The active faults are considered very dangerous in the geology of any area and in NWFP these faults exist in Chitral, Ghizer, Gilgit, Skardu, Diamer, Kohistan, Swat, Lower Dir, Bajur, Mahmand, Kurram, Khyber, Adem Khail, Noshara and Kohat. Whereas in Punjab the active faults are located in Mianwali, D.G Khan and Chkwal. In Balochistan active faults are located in Kohlu, Qila Saifullah, Tank, Musa Khail, Sibi, Quetta, Ziarat, Pashin, Kuch, Gwader, Chagi, Qila Abdullah, Awaran. All the districts with active faults are relatively more dangerous than the other districts in Pakistan. The GIS produced by the geological and seismological data sets give the information that the capital of Pakistan lies on the thrust fault .The districts with these fault lines are Kashmir, Bagh, Muzaffarabad, Kharpur, Haripur, North Wazistan, Zhob, Qilasafullah, Barkhan, Loralai, Kohlu, Khuzdar, Kharan and Chaghi.

Reverse fault is also a very dangerous fault in geological studies. These districts of Pakistan e.g. Adam khail Orakzi, Hangu, Tank, South Wazeeristan,

Zhob, Musa Khail, Qila saifulah, Pishin, Qila Abduallah, Sibi, Kahal, Panjgur, Khuran, Kuch and Awaran are on the reverse fault.

Strike Slip fault is only located in Khuran, Pangur, Kuch and Kalat districts of Balochistan and mostly undifferentiated faults are in Balochistan. The linements mostly exist in Larkana, Naushahro Feroz. Shakar Pur, Naseerabad, Nawab Shah, Lasbela, Haiderabad, Thatta, Karachi, Mianwali, Chakwal, Mandi Bahauddin, Gujrat, Haripur,Laki Marwat, D. I. Khan, Buner, Shangla, Butgram and Swabi.

The above discussion shows that NWFP and Balochistan are seismically risky regions and main centers of earthquakes in Pakistan.

This GIS of geological fault locations is the complete database of the fault lines in Pakistan. We can use this GIS database to analyze seismic activity and to estimate the seismic risk. This map also shows that most of the areas in Pakistan are seismically active regions and the mountain ranges (e.g. Himalayas, Karakorum and Hindu Kush) in the northern part of the country also support this opinion.

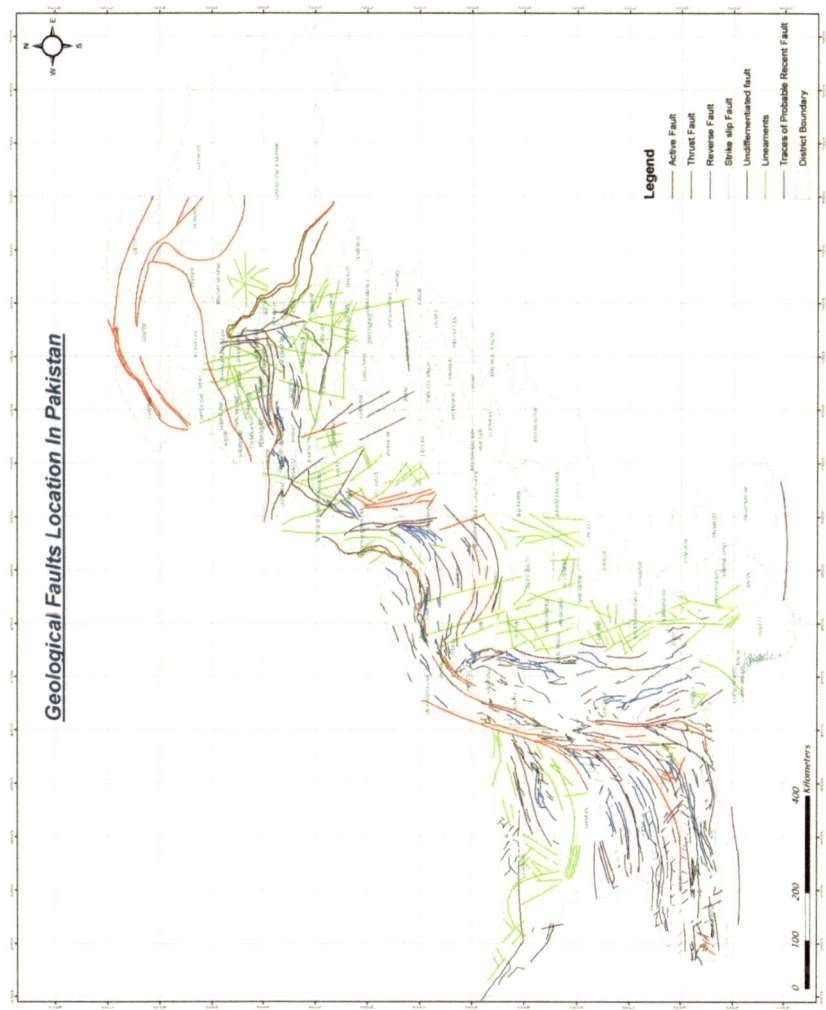

Figure 6. 1   Geological faults location map of Pakistan

6.2 Analysis of P_n Velocity Model

The map shown in Figure 6.2 indicates the different selected locations of the country where we have calculated the value of P_n velocity. Different colors in the map legend describe the provinces of Pakistan and special symbol "×" shows locations where P_n velocity has been calculated. Most locations are in NWFP and Balochistan province because most of the earthquake epicenters are in these provinces. Some points are from Gilgit-Baltistan, and some of them have been taken from the mutual border area of NWFP, Balochistan and Punjab. The map also shows the geographic locations of these points.

By the use of these points, we have generated the P_n velocity model. For this model we get the P_n velocity values all over the study area. This P_n velocity model is shown in Figure 6.3

This model shows the variations in the values of P_n velocity. The legend gives an idea that the maximum value of the P_n velocity is 7.34 Km/s and minimum value of P_n velocity is 4.63 Km/s. In costal areas of Sindh, value of P_n velocity is low whereas toward the Sindh Punjab border the value of P_n velocity is high and reaches the maximum. Similarly in NWFP, tribal areas of Pakistan, southern Punjab and Chaghi the value of P_n velocity is high.

According to the laws of refraction of wave, velocity of wave is high in rare medium and low in denser medium. So values of P_n velocity describe the structure of Basaltic layer of the Earth's crust. We have analyzed that the regions where P_n velocity is high, the rocks of those regions are less dense and where P_n velocity is low the rocks of those regions are denser. So this variation in P_n velocity describes the basaltic layer structure of Earth's crust.

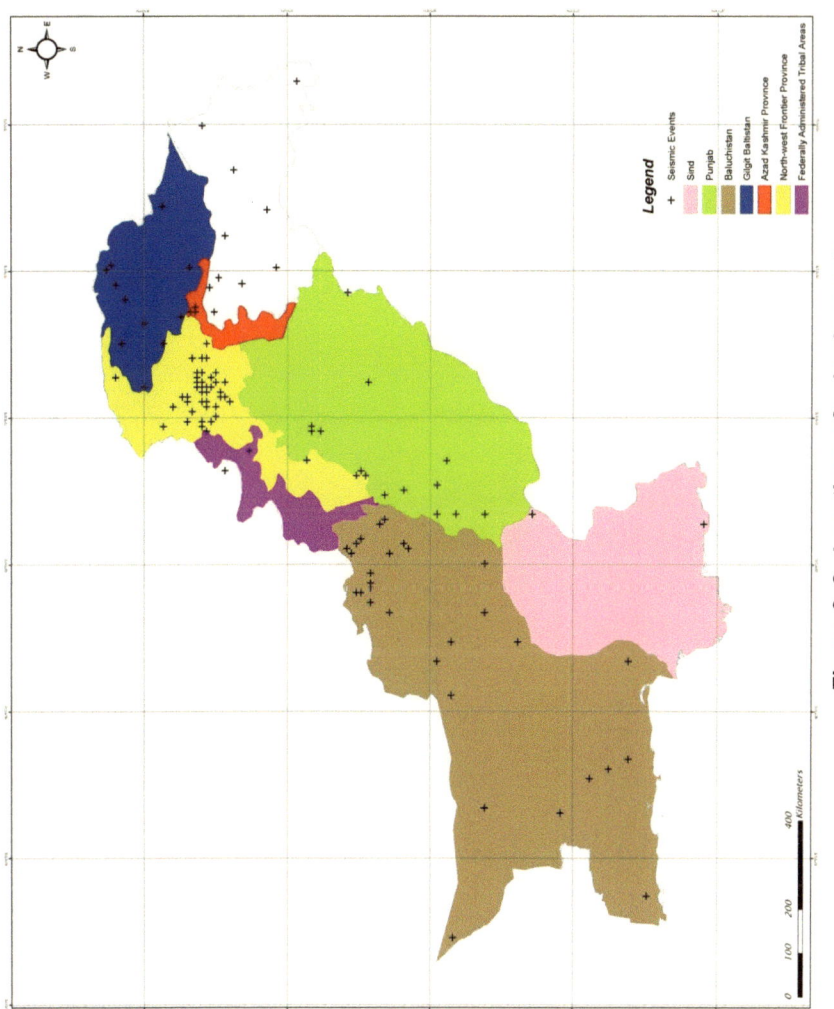

Figure 6. 2   Locations of seismic events

Legend
+ Seismic Events
Sind
Punjab
Baluchistan
Gilgit Baltistan
Azad Kashmir Province
North-west Frontier Province
Federally Administered Tribal Areas

0   100   200   400
Kilometers

39

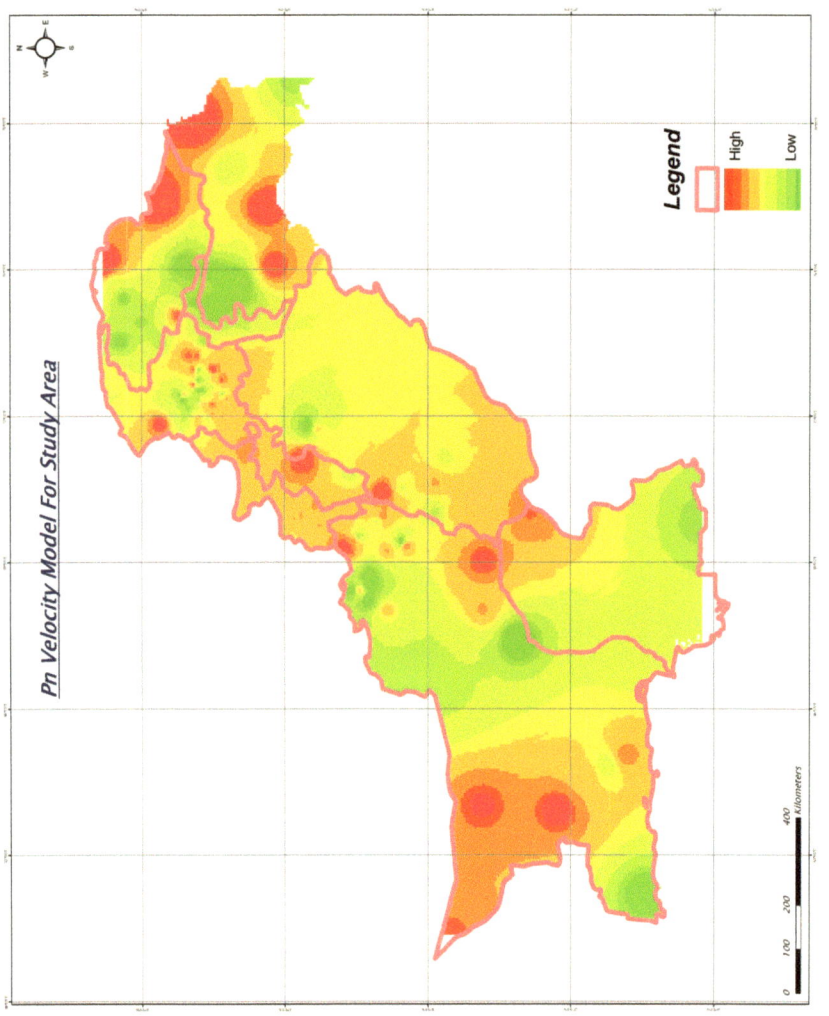

Figure 6. 3   Pn velocity model for study area

6.3 Analysis of P_g Velocity Model

P_g velocity location map in Figure 6.2 shows the points where P_g velocity values have been calculated. Every location has a specific value of P_g velocity. In the map provinces of Pakistan are represented by using different colors and locations of P_g velocity are shown by specific symbol "+" (see legend). P_g velocity is calculated by seismograph which is recorded after earthquake. Majority of points have been taken from NWFP, Balochistan, Gilgit-Baltistan and Punjab province and only two points have been taken from Sindh. Major seismic active areas exist in NWFP, Balochistan, Gilgit Baltistan and Punjab province. The point values of P_g velocity are important for us. These values are used in P_g velocity model for calculating P_g velocity for the entire study area. This model helps us study the structure of granitic layer. That's why we are very conscious in selecting the locations where P_g velocity is calculated. This map shows the spatial locations of P_g velocity.

P_g velocity model map shown in figure 6.4 represents the variation in the P_g velocity all over the study area .This variation is shown in map by different colors (see legend). The P_g velocity ranges from 6.08 Km/s (minimum) to 9.8 Km/s (maximum).

Red color in the map represents the highest velocity areas and Blue color has lowest velocity areas. It is clear in map that Gilgit-Bultistan and southern Balochistan fall in high P_g velocity zone. In Sindh, P_g velocity is moderate whereas in Punjab and central Balochistan region variation in P_g velocity is very clear. Similarly in different parts of NWFP P_g velocity shows abrupt variation.

We are applying law of refraction for the description of granitic layer structure of the Earth's crust. When the wave is propagating in denser medium its velocity is slower than the wave propagate in rare medium. According to this law we can say that the areas have rare density of rocks where the value of P_g velocity is found high and the areas having denser rocks, the values of pg velocity is found low. We have explained the granitic layer structure of the Earth's crust by the help of the P_g velocity mode

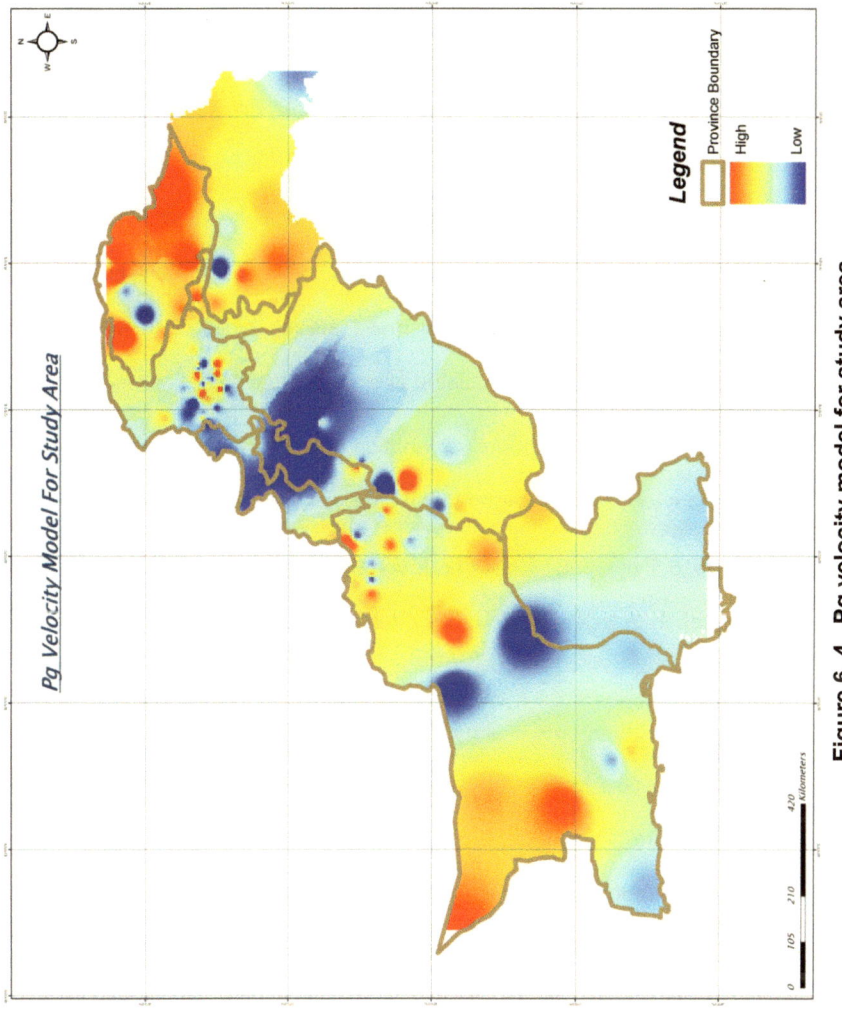

Figure 6. 4   Pg velocity model for study area

43

6.4   Analysis of Moho Depth Variation

Moho depth provides information about the thickness of the Earth's crust. The map shown in Figure 6.2 indicates the location of points where moho depth has been calculated. The symbol "⊗" represents those particular areas in map. The legend should be consulted for more details about the themes and colors presented in the map. Moho depth has been calculated at some specific locations of the country. Most of the moho depth points have been taken from the regions of Kashmir, Punjab, NWFP and Balochistan. Because major seismic activity occurs in these areas, therefore in these areas the moho depth values are calculated at specific locations.

We have to show the counties variation of moho depth all over the study area therefore we have converted these point values into surface values.

Figure 6.5 presents the surface variation of moho depth using color ramp mentioned in legend of the map (from Blue to Red). The maximum value of moho depth is 79.35 Km and minimum value is 23.40Km. The moho depth values of northern areas like Himalaya range, Karakaram range etc. are more than the central and southern regions of the study area. Conversely Kashmir and northern Punjab contains relatively less values than the northern part of study area. Southern Punjab and central Balochistan have moderate value of moho depth. The lower most part of study area is costal zone having lowest values of moho depth. Due to non-availability of moho depth data in the Badin district the area is left blank

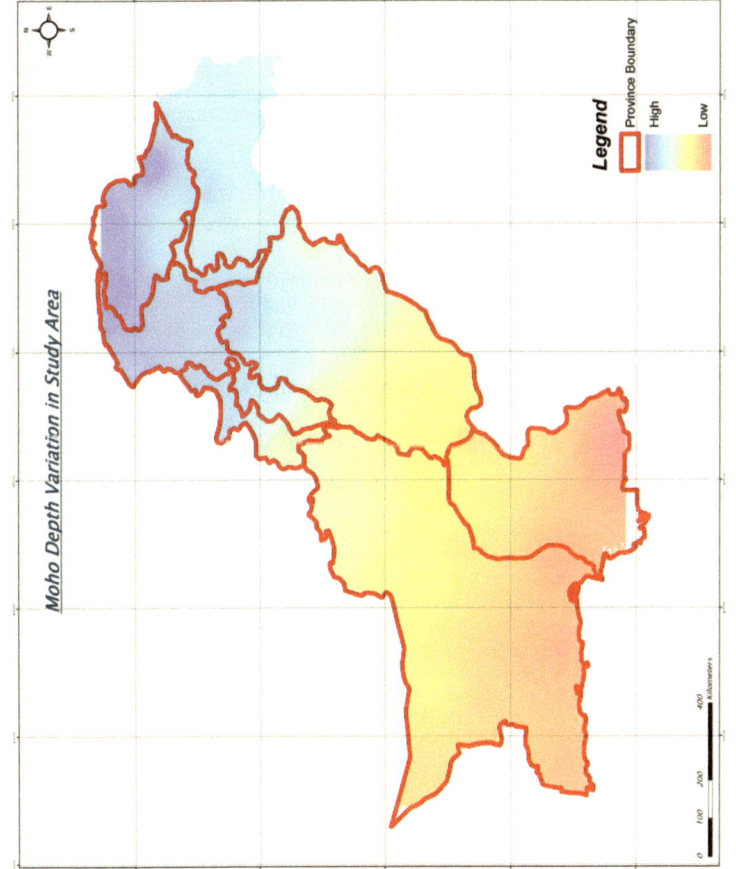

Figure 6. 5   Moho depth variation in study area

6.5 Seismotectonic GIS of Pakistan

The maps shown in Figures 6.6, 6.7, and 6.8 are seismotectonic GIS of Pakistan. The data used to produce the map, shown in figure 6.6, is geographical, geological and P_n velocity. In the other two maps P_g velocity and moho depth data have been used with geographical and geological data sets of study area. In all three maps variation in seismological data sets are represented by different colors (see map legend).Scale bar in each map is for the manual measurement of fault lines and corresponding areas. These maps are in accordance with the international standard. Variations in the seismological parameters have already been discussed in sections 6.3, 6.4 and 6.5. But now we are analyzing these parameters with fault lines. Almost all types of fault lines exist in NWFP and tribal areas of Pakistan, but active faults dominate in these regions. The value of moho depth and P_g velocity is high whereas P_n velocity is low in NWFP and tribal areas of Pakistan. According to these results, we can say that earthquakes from these regions have epicenters in deep but due to high velocity of P_g more ground shaking is expected. In Azad Kashmir the values of P_g velocity and P_n velocity are high but moho depth relatively less. This behavior of seismological parameters with fault lines shows that earthquakes from this region have shallow epicenters. Thus the Intensity of earthquakes is high in this region.

Moho depth is uniform from Pakpatan, Sahiwal, Tobatake singh, Jhang, Khoshab, Bannu and Kurram to Kashmir, but P_g and P_n velocities are not uniform. Because only in Kashmir region active and thrust fault exist and other districts have almost plane topography. Similar condition can be seen in districts of southern Punjab, interior Sindh and central Balochistan where moho depth is uniform whereas P_g and P_n velocities are not. These districts have plane topography but have different types of rocks.

In costal areas of Pakistan values of moho depth are almost uniform because topography is uniform. But the important analysis is that values of P_g and P_n velocities are also uniform. This shows that there is no significant variation in the tectonic structure of the costal areas in Pakistan. The combination of the seismological, geological and geographical data sets produced the complete seismotectonic GIS of Pakistan.

The map shown in Figure 6.9 is seismic hazard map of Pakistan. This map helps us study the seismically hazardous areas in the country. Red color in the map represents the highly seismic areas and green color low seismic areas. This map has been produced by applying regression technique on the seismological parameters. The regression equation and regression value of seismological parameters are given below:

$$y = 0.0231x + 7.06 ....(1)$$
$$R^2 = 0.9447$$

Where y = hazard intensity

x = contribution of seismological parameters

R^2 = regression

According to the map shown in Figure 6.9 these districts of Pakistan Abbotabad, Bannu, Batgram, Buner, Charsadda, Chitral, D.I. Khan, Hangu,

Haripur, Karak, Kohat, Kohistan, Lakki, Lower Dir, Malakand, Mansehra, Mardan, Nowshehra, Peshawar, Shangla, Swabi, Swat, Tank, Upper Dir, Diamer, Ghanche, Ghizer, Gilgit, Skardu, Islamabad Adam Khel, Bajur, Khyber, Kurram, Mahmand, North Waziristan, Orakzai, South Waziristan ,Bagh, Kotli, Mirpur, Muzaffarabad ,Poonch, Bhakkar, Layyah, Loralai and Kalat are highly seismic zones. On the other hand districts of Sindh province, southern Blochistan, southern Punjab and middle Punjab are lying in low seismic areas. The high value of R^2 indicates that seismological parameters are highly correlated for analyzed data set.

7.0 Summary

Pakistan and their neighboring counties contain all recognized kinds of most important plate boundaries in its region as well as important dynamic intra-plate twist. Understanding the tectonics in this multifaceted area has been delayed by a relative lack of data and the difficulty of the geologic and tectonic troubles. Even with the raise in the quantity of data in the past few years, the complications of the area need a multidirectional approach to be aware of the geology and tectonics. Thus, in order to grip huge, multidirectional data sets with varying superiority and resolution, we have take on a Geographic Information System (GIS) approach for erection of a versatile database to seem at these troubles in a comprehensive and exceptional way. In this study, we nearby innovative set maps of surficial tectonic features and deepness to the Moho, P_n velocities and P_g velocities for the Pakistan and explain a cross-section means to work with data in a GIS format.

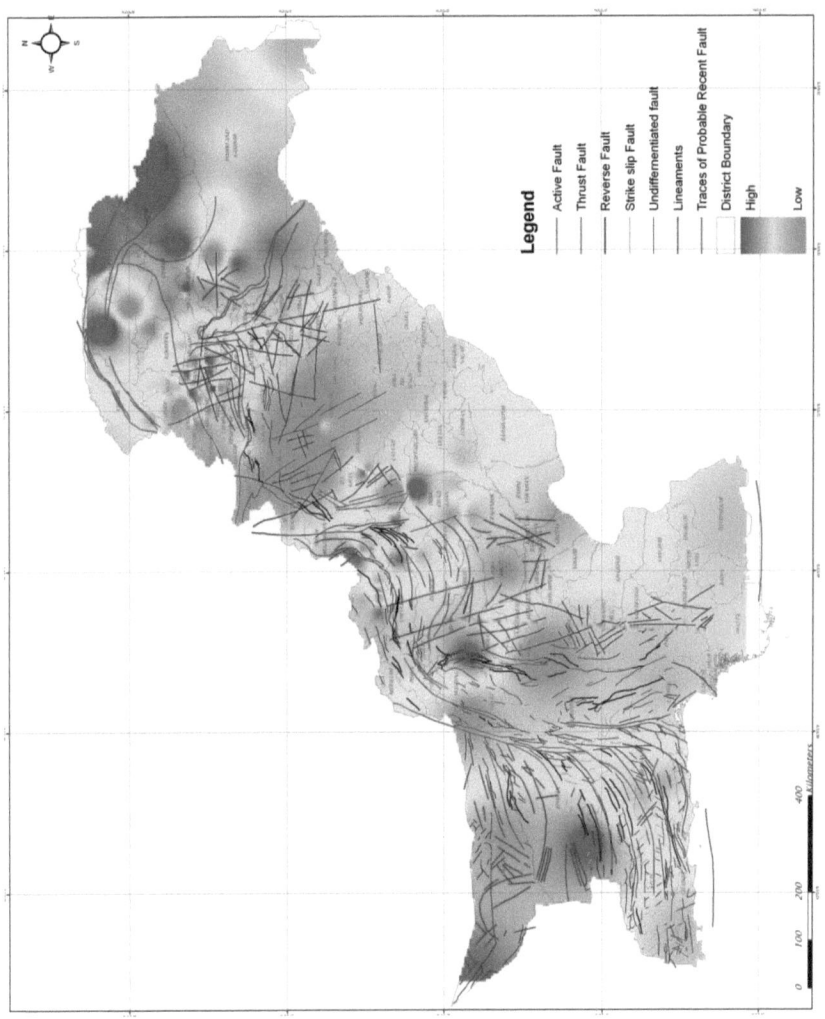

Figure 6. 6 Geological faults in Pakistan and Pn with districts

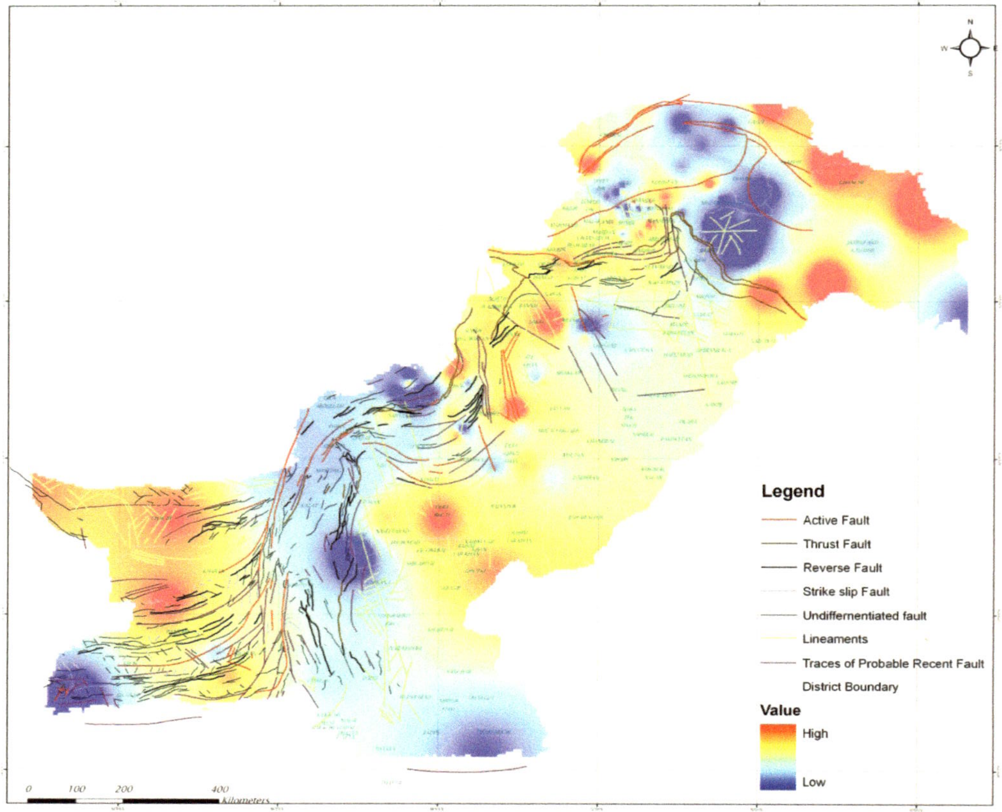

Figure 6. 7 Geological faults in Pakistan and Pg with districts

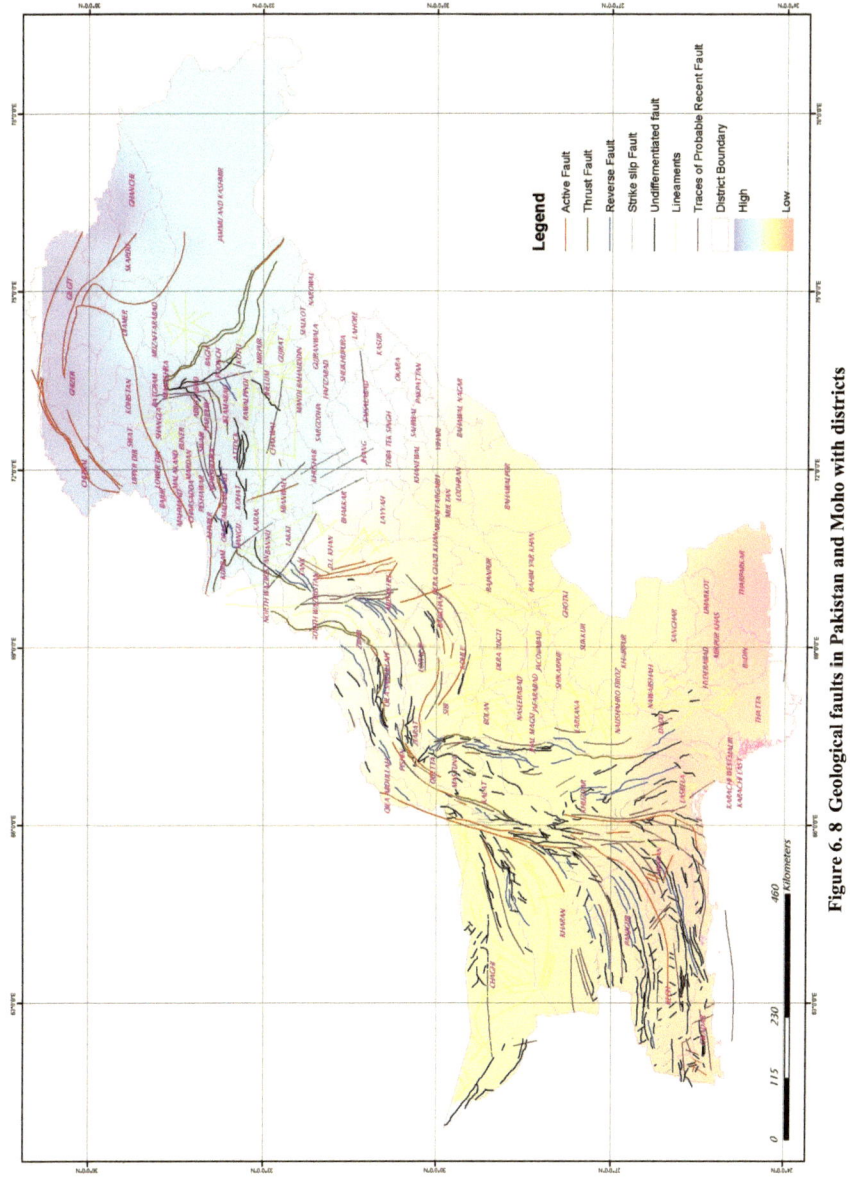

Figure 6. 8 Geological faults in Pakistan and Moho with districts

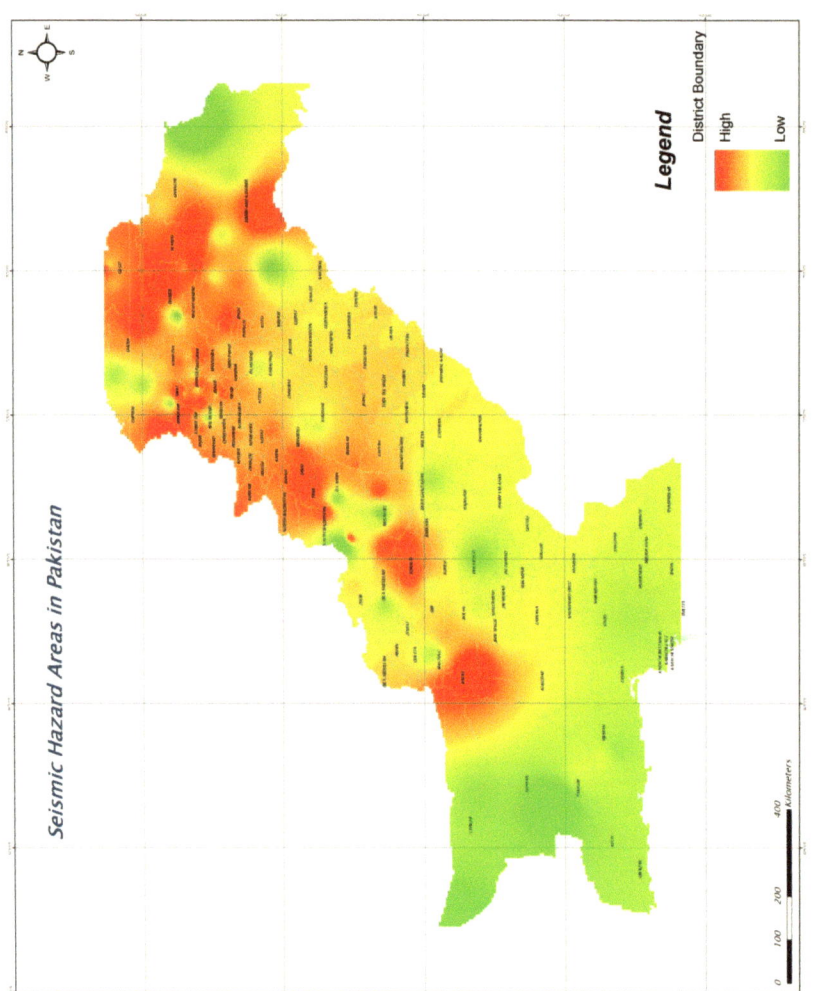

Figure 6.9   Sismic Hazard Areas in Pakistan

References

Abarca, M.A.A., 2006. Lineament extraction from digital terrain models: case study San Antonio del Sur area, south-eastern Cuba. M.S. Italy: International Institute for Aerospace Survey and Earth Observation (ITC).

Ali, S.A., & Pirasteh, S., 2004. Geological applications of Landsat Enhanced Thematic Mapper (ETM) data and Geographic Information System (GIS): mapping and structural interpretation in south-west Iran, Zagros Structural Belt. *International Journal of Remote Sensing*, 25(21), pp.4715-4727.

Armbruster, J.G., Seeber, L. & Jacob, K.K., 1978.The northwest termination of the Himalayan mountain front: active tectonics from micro earthquakes. *J. Geophys*. Res., 83, pp.269-282.

Auden, J. B., 1938. Geological Results. In: E. SHIPTON, ed. *The Shaksgam Expedition 1938. Geographical Journal,* 91, pp.335-336.

Barazangi, M., Fielding, E. J., Isacks, B., & Seber, D., 1996. Geophysical and geological databases and CTBT monitoring: a case study of the Middle East, In Monitoring a Comprehensive Test Ban Treaty. In: E.S. Husebye & A.M. Dainty, eds. Kluwer Academic Publishers: The Netherlands, pp.197-224.

Burget, J. P. et al., 1996. Structural evidence for backsliding of the Kohistan arc in the couisional system of NW Pakistan. *Journal of Geology,* 24, pp.739-742.

Camp, V. E., & Roobol, M. J., 1992. Upwelling asthenosphere beneath western Arabia and its regional implications, *J. Geophy.* Res., 97, pp.15255-15271.

Chaimov, T. et al., 1990. Crustal shortening in the Palmyride fold belt, Syria, and implications for movement along the Dead Sea Fault system, *Tectonics,* 9, pp.1369-1386.

Clark, C.D., Wilson, C., 1994. Spatial analysis of lineaments. *Computers and Geosciences*, 20, pp.1237-1258.

Davis, G.H., 1984. *Structural Geology of rocks and regions*. New York: Wiley, p.475.

Deslo, A., 1930. Geological work of the Italian expedition to the Karakoram. *Geographical Journal,* 75,pp. 402-411.

Dewey, J. F., & Sengor, A.M.C., 1979. Aegean and surrounding regions: complex multiple and continuum tectonics in a convergent zone. *Geol. Soc. Am. Bull.* 90, pp.84-92.

Dhakal, A.S., Amada, T. & Aniya, M. 2000. *Landslide hazard mapping and its evaluation using GIS: an investigation of sampling schemes for a grid-cell based quantitative method*. Photogrammetric Engg. & Remote Sensing, 66(8), pp.981–989.

Edwards, R.A., Minshull, T.A., & White, R.S, 2000. Extension across the Indian-Arabian plate boundary: the Murray Ridge. *Geophysical Journal International*, 142, pp.461-477.

Eide, A.L., Omre, H., & Ursin, B., 2002. Prediction of reservoir variables based on seismic data and well observations. *Journal of the American Statistical Association*, 97, pp.18-28.

Ghalib, H. A. A., 1992. Seismic velocity structure and attenuation of the Arabian plate, Ph.D. St. Louis, USA: Louis University, p.314.

Hawkins, D.M., 1974. The detection of errors in multivariate data, using Principal Components. *Journal of the American Statistical Association*, 69, pp.340-344.

Hearn T M., 1999. Uppermost mantle velocities and anisotropy beneath Europe. *J Geophys* Res, 104(B7), pp.15123-15139.

Jadoon, I.A.K., Lawrence, R.D., & Khan, S.H, 1994. Mari-Bugti pop-up zone in central Sulaiman fold belt, Pakistan. *Journal of Structural Geology*, 16, pp.147-156.

Jansson, K.N., & Glasser, N.F., 2005. Using Landsat 7 ETM+ imagory and Digital Terrain Models for mapping glacial lineaments on former ice sheet beds. *International Journal of Remote Sensing*, 26,pp.3931-3941.

Johnson, N.M. et al., 1985. Paleomagnetic chronology, fluvial processes, and tectonic implications of the Siwalik deposits near Chinji Village, Pakistan. *Journal of Geology*, 93, pp.27-40.

Leech, D.P., Treloar, P.J., Lucas, N.S., & Grocott, J., 2003. Landsat TM analysis of fracture patterns: a case study from the Coastal Cordillera of northern Chile. *International Journals of Remote Sensing*, 24, pp.3709-3726.

Le Pichon, X., & J. Francheteau, 1978. A plate-tectonic analysis of the Red Sea - Gulf of Aden area. *Tectonophysics*, 46, pp.369-406.

Masoud, A., & Koike, K., 2006. Tectonic architecture through Landsat-7 ETM+/SRTM DEM-derived lineaments and relationship to the hydrogeologic setting in Siwa. *Journal of African Earth Sciences*, 45, pp. 467-477.

Mostafa, M.E., & Bishta, Z.A., 2005. Significance of lineament patterns in rock unit classification and designation: a pilot study on the Gharib-Dara area, northern eastern desert, Egypt. *International Journal of Remote Sensing*, 26, pp.1463-1475.

Nama, E.E., 2004. Lineament detection on Mount Cameroon during the 1999 volcanic eruptions using Landsat ETM. *International Journal of Remote Sensing*, 25, pp.501-510.

North, H.C., & Pairman, D., 2001. Edge detection as a starting point for remote sensing scene interpretation. *IEEE*, pp.2994-2997.

O'Learly, D.W., Friedman, J.D., & Pohn, H.A., 1976. Lineament, linear, lineation: some proposed new standards for old terms. *Bulletin of the Geological Society of America*, 87, pp.463-1469.

Petterson, M. G., Jan, M. Q. & Sullivan, M., 1996. A re-evaluation of the stratigraphy and evolution of the Kohistan arc sequence, Pakistan Himalaya: implications for magrnatic and tectonic arc-building processes. *Journal of the Geological Society, London*, 153, pp.677-680.

Pognante, U., & Spencer, D. A., 1991. First record of eclogites from the Himalayan belt, Kaghan Valley, Northern Pakistan. *European Journal of Minerology*, 3, pp.613-618.

Sandvol, E., Seber, D., Barazangi, M., 1996. Single-station receiver function inversions in the Middle East and North Africa: a grid search approach, *AGU fall meeting*. Abstract only.

Searle, M. P., 1986. Structural evolution and sequence of thrusting in the High Himalaya, Tibetan- Tethys and Indus suture zones of Zanskar and Ladakh, western Himalaya. *Journal of Structural Geology*, 8, pp.923-936.

Sengor, A.M.C., Gorur, N., & Saroglu, F., 1985. Strike-slip faulting and related basin formation in zones of tectonic escape: Turkey as a case study. In: K.T. Biddle & N. Christe-Blick, eds. *Strike-slip deformation, basin formation and sedimentation, Soc. Economic Paleontologist and mineralogists,* Special Publication, 37, pp.227-265.

Smith, K. A., Churilov, L., Siew, E. G., & Wassertheil, J. 2005. Towards data-driven acute in-patient classification schemes: a hospital management perspective. *Central European Journal of Operations Research*, 13(4), pp.365-392.

Tahirkheli, R.A.K., Mattauer, M., Proust, F. & Tapponier, P., 1979. The India-Eurasia suture zone in northern Pakistan: some new data for interpretation at plate scale. In: A. Farah, K.A. DeJong, eds. *Geodynamics of Pakistan,* pp.125-130.

Vince, K. J., & Treloar, P. J., 1996. Miocene, northvergent extensional displacements along the Main Mantle Thrust, NW Himalaya, Pakistan. *Journal of the Geological Society, London,* 153, pp.677-680.

Yeats, R.S. & Lawrence, R.D., 1984. Tectonics of the Himalayan thrust belt in northern Pakistan. In: B.U. Haq , & J.D. Milliman, eds. *Marine geology and oceanography of Arabian Sea and coastal Pakistan,* pp.177-200.

Yun, S.H., & Moon, M.W., 2001. Lineament extraction from DEM using drainage network. School of Environmental Sciences, Seoul National University, Kwanak. *IEEE,* pp.2337-233

Appendix A: Training and Experience

The following are different training events attended during the research work.

Departmental Experience

It was a very good experience to study at the Department of Meteorology and here I have learnt advanced techniques of Remote Sensing and Geographical Information System (GIS).

Workshops Attended

1. Participated in the five days National Workshop on "Seismicity, Seismotectonic, Neotectonics and Earthquake Forcasting in Northern Pakistan" held on July 20-24, 2009 at National Centre for Physics (NCP), Quaid-i-Azam University, Islamabad.

2. A three day workshop on "Analysis of research data using SPSS" was attended from 23-25 April 2009 at COMSATS Institute of Information Technology.

Seminars Attended

"GEOTECHICAL DISASTER MITIGATION" held on the 9th June, 2009 at Auditorium, National Highway Authority, Islamabad. The seminar was organized by University of Tokyo, Japan International Cooperation Agency, and Geological Survey of Pakistan in cooperation of National Highway Authority ,Engineers without Borders, Japan, National Disaster Management Authority, and Earthquake Rehabilitation & Reconstruction Authority.

APPENDIX B: Districts names and their Provinces

Sr. No.	Name	Province
1	Abbotabad	NWFP
2	Adam khel	Fata
3	Attock	Punjab
4	Awaran	Balochistan
5	Badin	Sindh
6	Bagh	Azad jammu and kashmir
7	Bahawal nagar	Punjab
8	Bahawalpur	Punjab
9	Bajur	Fata
10	Bannu	NWFP
11	Barkhan	Balochistan
12	Batgram	Nwfp
13	Bhakkar	Punjab
14	Bolan	Balochistan
15	Buner	Nwfp
16	Chaghi	Balochistan
17	Chakwal	Punjab
18	Charsadda	NWFP
19	Chitral	NWFP
20	D.i. Khan	NWFP
21	Dadu	Sindh
22	Dera bugti	Balochistan
23	Dera ghazi khan	Punjab
24	Diamer	Northern areas
25	Faisalabad	Punjab
26	Ghanche	Northern areas
27	Ghizer	Northern areas
28	Ghotki	Sindh
29	Gilgit	Northern areas

Sr. No.	Name	Province
30	Gujranwala	Punjab
31	Gujrat	Punjab
32	Gwadar	Balochistan
33	Hafizabad	Punjab
34	Hangu	NWFP
35	Haripur	NWFP
36	Hyderabad	Sindh
37	Islamabad	Federal capital
38	Jacobabad	Sindh
39	Jafarabad	Balochistan
40	Jammu and kashmir	Disputed territory
41	Jhal magsi	Balochistan
42	Jhang	Punjab
43	Jhelum	Punjab
44	Kalat	Balochistan
45	Karachi central	Sindh
46	Karachi east	Sindh
47	Karachi south	Sindh
48	Karachi west	Sindh
49	Karak	NWFP
50	Kasur	Punjab
51	Kech	Balochistan
52	Khairpur	Sindh
53	Khanewal	Punjab
54	Kharan	Balochistan
55	Khushab	Punjab
56	Khuzdar	Balochistan
57	Khyber	Fata
58	Kohat	Nwfp
59	Kohistan	Nwfp
60	Kohlu	Balochistan

Sr. No.	Name	Province
61	Kotli	Azad jammu and kashmir
62	Kurram	Fata
63	Lahore	Punjab
64	Lakki	NWFP
65	Larkana	Sindh
66	Lasbela	Balochistan
67	Layyah	Punjab
68	Lodhran	Punjab
69	Loralai	Balochistan
70	Lower dir	NWFP
71	Mahmand	Fata
72	Malakand	Nwfp
73	Malir	Sindh
74	Mandi bahauddin	Punjab
75	Mansehra	NWFP
76	Mardan	NWFP
77	Mastung	Balochistan
78	Mianwali	Punjab
79	Mirpur	Azad jammu and kashmir
80	Mirpur khas	Sindh
81	Multan	Punjab
82	Musakhel	Balochistan
83	Muzaffarabad	Azad jammu and kashmir
84	Muzaffargarh	Punjab
85	Narowal	Punjab
86	Naseerabad	Balochistan
87	Naushahro firoz	Sindh
88	Nawabshah	Sindh
89	North waziristan	Fata
90	Nowshehra	NWFP
91	Okara	Punjab

Sr. No.	Name	Province
92	Orakzai	Fata
93	Pakpattan	Punjab
94	Panjgur	Balochistan
95	Peshawar	Nwfp
96	Pishin	Balochistan
97	Poonch	Azad jammu and kashmir
98	Qila abdullah	Balochistan
99	Qila saifullah	Balochistan
100	Quetta	Balochistan
101	Rahim yar khan	Punjab
104	Rajanpur	Punjab
105	Rawalpindi	Punjab
106	Sahiwal	Punjab
107	Sanghar	Sindh
108	Sargodha	Punjab
109	Shangla	Nwfp
110	Sheikhupura	Punjab
111	Shikarpur	Sindh
112	Sialkot	Punjab
113	Sibi	Balochistan
114	Skardu	Northern areas
115	South waziristan	Fata
116	Sukkur	Sindh
117	Swabi	Nwfp
118	Swat	NWFP
119	Tank	NWFP
120	Tharparkar	Sindh
121	Thatta	Sindh
122	Toba tek singh	Punjab
123	Umarkot	Sindh
124	Upper dir	NWFP

APPENDIX C: Seismological Parameters

Sr. No.	Date	Time UTC	Longitude	Latitude	Magnitude	P_g Velocity (Km/s)	Moho Depth (Km)	P_n Velocity (Km/s)
1	24-01-1983	1137	73.94	36.00	5.8	7.20	68.52	5.40
2	24-05-1986	1942	62.24	25.47	5.8	7.90	27.63	5.20
3	13-05-1987	1243	71.94	35.10	6.6	7.50	65.86	5.60
4	8-09-1989	2008	72.44	35.20	5.1	7.83	66.54	5.30
5	9-09-1989	2240	71.13	29.65	5.6	7.98	32.58	5.90
6	29-07-1990	923	72.84	34.60	6.2	8.13	59.68	5.81
7	17-01-1991	335	72.64	34.80	5.7	8.28	61.35	5.92
8	09-04-1991	931	72.34	34.70	5.9	8.43	63.58	6.03
9	18-10-1992	808	74.58	31.73	5.3	8.58	40.52	6.14
10	14-10-1993	33	70.03	29.45	5.2	8.73	35.36	6.25
11	16-11-1994	1422	72.24	34.70	5.3	8.88	58.60	6.36
12	17-11-1994	2108	69.23	30.85	6.3	9.03	37.65	6.47
13	27-11-1994	2131	69.43	31.55	6	7.50	36.87	6.58
14	06-11-1994	759	69.33	30.45	5.5	7.83	35.69	6.69
15	09-08-1995	1414	72.74	34.30	6.5	7.98	62.36	6.80

Sr. No.	Date	Time UTC	Longitude	Latitude	Magnitude	P_g Velocity (Km/s)	Moho Depth (Km)	P_n Velocity (Km/s)
16	21-08-1995	1218	73.24	34.80	6.1	8.26	61.58	6.91
17	20-10-1995	1051	75.04	36.80	5.4	8.50	69.79	7.02
18	06-03-1996	2036	71.84	35.60	5.5	8.74	68.65	7.13
19	05-03-1996	1409	71.14	32.60	5.5	6.08	55.36	7.24
20	05-03-1996	2154	70.43	30.95	5.7	6.32	45.32	7.35
21	15-11-1996	1646	72.94	34.80	6.2	6.56	63.84	5.26
22	06-03-1997	710	72.14	35.00	6	6.81	66.32	5.37
23	13-03-1997	2311	72.44	34.30	6.2	7.05	67.95	5.48
24	24-09-1998	1752	66.33	29.55	6	7.29	34.65	5.59
25	29-10-1999	2254	72.64	34.60	6.2	7.50	62.35	5.70
26	22-02-2000	222	72.64	34.70	6	7.83	63.25	5.81
27	25-02-2001	2043	73.24	34.70	6	7.98	64.15	5.92
28	05-04-2004	705	72.54	34.40	6.1	8.26	65.05	6.03
29	14-04-2004	1208	71.74	34.70	6.9	7.18	66.95	6.14
30	13-05-2004	1457	70.94	34.30	6	7.39	67.85	6.25
31	15-08-2004	2006	70.93	31.45	6	7.60	37.54	6.36
32	18-09-2004	1147	72.34	34.20	5.8	7.83	55.32	6.47
33	20-10-2004	2124	72.74	34.80	6.8	7.98	58.65	4.82

Sr. No.	Date	Time UTC	Longitude	Latitude	Magnitude	P_g Velocity (Km/s)	Moho Depth (Km)	P_n Velocity (Km/s)
34	30-11-2004	1413	72.34	35.10	5	8.26	63.58	4.93
35	02-03-2005	401	69.43	30.55	5.5	8.45	40.32	5.04
36	03-04-2005	148	72.54	34.70	5.5	8.66	63.52	5.15
37	05-02-2005	2304	74.44	36.40	5.9	7.83	69.98	5.26
38	03-03-2005	221	73.54	34.70	5.2	7.98	68.35	5.37
39	08-10-2005	351	74.68	34.63	7.5	8.26	58.36	5.48
40	11-08-2005	1046	74.18	34.93	6.2	8.45	59.35	5.59
41	21-12-2005	2024	74.18	35.03	5.5	8.66	60.34	5.70
42	15/07/2005	1928	74.18	34.53	5	8.87	61.33	4.82
43	14-05-2005	143	74.75	33.94	5.2	9.08	62.32	4.93
44	04-11-2005	2149	74.28	34.93	5	9.29	63.31	5.04
45	16-08-2005	1246	75.08	35.06	5	9.50	64.30	5.15
46	08-07-2005	1722	73.53	36.47	5.9	9.71	70.32	5.26
47	28-10-2005	335	72.34	34.80	6	9.92	66.28	5.37
48	19-06-2005	37	73.54	35.60	5.1	8.87	67.27	5.48
49	04-10-2005	201	70.83	31.55	5.3	9.08	45.65	5.59
50	21-09-2006	720	67.43	29.55	5.1	9.29	47.32	5.70
51	25-09-2006	216	74.74	36.60	5.1	9.50	66.34	5.81

Sr. No.	Date	Time UTC	Longitude	Latitude	Magnitude	P_g Velocity (Km/s)	Moho Depth (Km)	P_n Velocity (Km/s)
52	30-09-2006	1354	70.53	30.55	5.1	9.71	41.36	5.92
53	06-08-2006	1157	72.74	34.90	5	9.92	65.32	6.03
54	16-09-2006	556	70.83	31.35	5.1	8.45	48.36	6.14
55	17-09-2006	1925	71.84	34.80	5	8.66	61.32	6.25
56	06-09-2006	2012	71.94	34.80	5.1	7.83	62.89	6.36
57	02-12-2007	20-57	72.44	35.10	5.2	7.98	64.46	6.47
58	0211-2007	204	71.34	33.80	5.2	8.26	66.03	6.58
59	04-04-2007	1828	72.24	34.50	5	8.45	67.60	6.69
60	03-05-2007	526	73.24	35.00	5	8.66	69.17	6.80
61	06-11-2007	458	72.94	34.90	5.3	8.87	70.74	6.91
62	18-01-2007	548	72.64	34.90	5.7	9.08	72.31	7.02
63	08-02-2007	255	75.14	36.70	5.4	9.29	73.88	7.13
64	08-08-2007	1800	72.94	34.50	5.3	9.50	67.60	7.24
65	09-09-2007	1553	76.34	35.62	6.5	9.71	69.17	7.35
66	18-12-2007	808	72.74	34.50	5.3	9.62	70.74	6.25
67	08-08-2007	129	72.44	34.40	5.6	9.76	72.31	6.36
68	19-10-2007	2233	68.43	31.45	5.2	8.58	34.65	6.47
69	09-11-2007	2310	68.43	31.55	6.5	8.73	35.78	4.82

Sr. No.	Date	Time UTC	Longitude	Latitude	Magnitude	Pg Velocity (Km/s)	Moho Depth (Km)	Pn Velocity (Km/s)
70	25-11-2007	1133	68.63	31.25	6.4	8.88	36.91	4.93
71	26-01-2007	247	68.23	31.25	5.1	9.03	38.04	5.04
72	06-01-2007	554	68.53	31.25	5.2	7.50	39.17	5.15
73	16-01-2007	2253	68.83	31.25	5.6	7.83	40.30	5.26
74	06-02-2008	2240	69.83	24.27	5.8	7.98	25.36	5.37
75	04-03-2008	338	72.84	34.90	5.7	7.65	67.68	6.36
76	15-09-2008	2023	72.04	34.50	6	7.47	65.23	6.47
77	05-10-2008	2313	72.54	34.80	5.8	7.30	62.78	4.82
78	25/10/2008	837	67.43	28.15	5	7.12	30.25	4.93
79	05-11-2008	349	74.88	34.43	5.4	6.94	49.36	5.04
80	10-09-2008	2128	71.84	32.50	5.4	6.76	50.23	5.15
81	06-10-2008	2343	71.74	32.50	5	7.50	51.10	5.26
82	06-11-2008	931	78.90	32.80	6	7.83	51.97	5.37
83	29-12-2008	808	69.53	31.45	7	7.98	34.56	5.48
84	19-03-2008	33	70.03	29.85	6	7.65	32.15	5.59
85	05-02-2008	1422	64.83	26.25	5.7	7.89	29.32	5.70
86	09-05-2008	2108	67.03	25.85	5.4	7.95	32.25	5.81
87	28-03-2008	2131	69.83	31.05	5.3	8.02	36.25	5.92

Sr. No.	Date	Time UTC	Longitude	Latitude	Magnitude	P$_g$ Velocity (Km/s)	Moho Depth (Km)	P$_n$ Velocity (Km/s)
88	11-05-2008	349	72.74	31.30	6.5	8.08	45.85	6.03
89	11-07-2008	1752	75.74	34.30	6	8.14	65.32	6.14
90	12-8-2008	2254	71.74	32.30	6	8.20	43.56	6.25
91	11-10-2009	222	64.63	26.65	6	8.26	30.25	6.36
92	13-10-2009	2043	68.03	28.85	7	8.32	32.74	6.47
93	19-01-2008	705	67.03	29.85	5.2	8.38	31.65	5.59
94	10-11-2008	1208	77.08	34.11	5	8.44	46.35	5.70
95	19-07-2008	1457	71.94	34.60	5	8.26	67.52	5.81
96	09-05-2008	2006	72.24	35.40	5.3	8.32	69.78	5.92
97	20-05-2008	1147	72.64	36.00	5.7	8.38	72.04	6.03
98	20-09-2008	2124	72.84	36.60	5.4	8.44	74.30	6.14
99	25-12-2008	1413	70.03	28.85	5.3	8.50	35.28	6.25
100	25-06-2008	401	68.03	30.85	6.5	8.56	38.65	6.36
101	28-10-2008	148	70.63	29.85	5.3	8.62	36.87	6.47
102	18-10-2008	1243	65.03	25.85	5.6	8.68	30.25	6.58
103	28-11-2008	33	70.03	27.85	5.2	8.74	33.46	6.69
104	28-05-2008	1422	64.03	28.85	6.5	8.80	36.67	6.80
105	29-10-2008	2108	69.03	28.85	6.4	8.87	39.88	6.91

Sr. No.	Date	Time UTC	Longitude	Latitude	Magnitude	P_g Velocity (Km/s)	Moho Depth (Km)	P_n Velocity (Km/s)
106	29-12-2008	2131	74.08	35.23	5.1	8.93	43.09	7.02
107	02-02-2009	349	75.08	33.23	5.2	8.99	46.30	7.13
108	12-03-2009	1752	69.93	30.95	5.6	9.05	49.51	6.58
109	03-01-2009	2254	61.38	29.51	5.8	9.11	52.72	6.69
110	04-10-2009	222	69.23	31.65	5.7	9.17	55.93	6.80
111	16-04-2009	148	63.93	27.25	6	9.23	59.14	6.91
112	20-02-2009	1243	69.33	31.75	5.8	9.29	62.35	7.02
113	20-03-2009	33	77.98	34.79	5	8.68	65.56	7.13
114	24-06-2009	1422	76.27	33.43	5.4	8.74	68.77	7.24